By **WAR DEPARTMENT**

GERMAN 88-MM ANTIAIRCRAFT GUN MATERIEL

TECHNICAL MANUAL

&

29 JUNE 1943

DISCLAIMER:

This manual is sold for historic research purposes
only, as an entertainment. It contains obsolete
information and is not intended to be used as part
of an actual operation or maintenance training
program. No book can substitute for proper training
by an authorized instructor.

★ RESTRICTED

TECHNICAL MANUAL }
No. E9-369A }

WAR DEPARTMENT
Washington, 29 June 1943

GERMAN 88-MM ANTIAIRCRAFT GUN MATERIEL

Prepared under the direction of the
Chief of Ordnance

CONTENTS

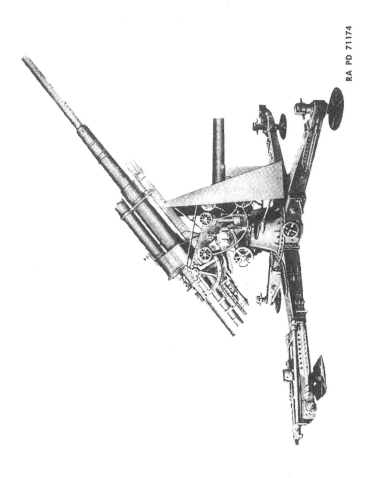

RA PD 71174

Figure 1 — German 88-mm Antiaircraft Gun — Firing Position

INTRODUCTION

RA PD 71175

Figure 2 — German 88-mm Antiaircraft Gun — Traveling Position

GERMAN 88-MM ANTIAIRCRAFT GUN MATERIEL

CHAPTER 1

INTRODUCTION

1. SCOPE.

a. This manual is published for the information and guidance of the using arms and services.

b. There is included as much technical information required for identification, use, and care of the German 88-mm antiaircraft gun as can be ascertained from printed matter and the materiel on hand. Corrections and additions to this manual will be published as the information becomes available.

c. In all cases where the nature of the repair, modification, or adjustment is beyond the scope or facilities of the unit or beyond the scope of this manual, the responsible ordnance service should be informed so that proper instructions may be issued.

2. CHARACTERISTICS.

a. The mount is a circular pedestal antiaircraft type suspended from two bogies when in traveling position. The mount is equipped with a data transmission indicator for antiaircraft fire. There are also provisions for installing a direct laying sight for antitank fire and a dial sight for indirect fire. A hand driven fuze setter is fitted to the left side of the top carriage. The traversing and elevating mechanism data transmission indicators and direct laying sight are on the right side.

b. Normally, the bottom carriage is in contact with the ground during firing and is stabilized by outriggers. There are four leveling jacks, one at each extremity of the outriggers for leveling the bottom carriage. The top carriage is leveled by two handwheels located 45 degrees from either side of the center line of the front outrigger on the bottom carriage. The leveling system has a range of 9 degrees. The front of the mount is protected by a flat shield of armor plate.

3. DATA.

a. Gun.

Type .... Tube and loose 3 section liner

Total weight .... 2,947 lb

Weight of removable components:

Breech ring .... 505.5 lb

INTRODUCTION

Outer tube	785 lb
Inner tube	805.5 lb
Liner (muzzle section)	600 lb
Liner (center section)	199 lb
Liner (breech section)	58 lb
Retaining rings	34 lb
Over-all length of tube	185 in. (470 cm)
Over-all length of gun and tube	194.1 in. (493.8 cm)
Length in calibers	56
Distance from center line of trunnions to breech face	6.5 in.
Travel of projectile in bore	157.4 in. (400 cm)
Volume of chamber	226 cu in.
Rated maximum powder pressure	33,000 lb per sq in. (approx.)
Muzzle velocity	2,690 ft per sec

Maximum range:

Horizontal	16,200 yd
Vertical	39,000 ft
Maximum effective ceiling	25,000 ft (at 70-deg elevation)

Rifling:

Length	157.4 in. (400 cm)
Direction	Right-hand
Twist	Increasing 1 turn in 45 calibers to 1 turn in 30 calibers
Number of grooves	32
Depth of grooves	0.0394 in. (1 mm)
Width of grooves	0.1969 in. (5 mm)
Width of lands	0.1181 in. (3 mm)
Type of breech mechanism	Semiautomatic horizontal sliding block
Rate of fire	15 rounds per min (practical rate at a mechanized target) 20 rounds per min (practical rate at an aerial target)

b. Recoil Mechanism.

Type	Independent liquid and hydropneumatic
Total weight	524 lb
Weight of recuperator cylinder	285 lb
Weight of recoil cylinder	239 lb
Weight of recoiling parts in recoil mechanism	108.5 lb
Total weight of recoiling parts (with gun and tube)	3,159 lb
Type of recoil	Control rod type with secondary control rod type regulating counterrecoil

GERMAN 88-MM ANTIAIRCRAFT GUN MATERIEL

Normal recoil:

    0-degree elevation .... 41.5 in. (105 cm)

    25-degree elevation .... 33.46 in. (85 cm)

    Maximum elevation .... 27.75 in. (70 cm)

Capacity of recoil cylinder .... 2.5 gal

Capacity of recuperator cylinder .... 4.5 gal

c. Mount.

Weight (less cannon and recoil mechanism)........................ 8,404 lb

Maximum elevation .... 85 deg

Maximum depression .... minus 3 deg

Traverse ....360 deg

Loading angles .... All angles

Height of trunnion above ground (firing position)...................... 5.2 ft

Height of working platform (firing)... 0.8 ft

Height of trunnion above working platform .... 4.4 ft

Leveling .... Pivots located 45 deg from either
                                  side of center line of front
                                  outrigger (total of 9 deg each)

Number of turns of handwheel to elevate from 0 to 85 degrees:

    High gear 42.5

    Low gear 85

Elevation for one turn of elevating handwheel:

    High gear .... 2 deg (35.4 mils)

    Low gear .... 1 deg (17.7 mils)

Number of turns of handwheel to traverse 360 degrees:

    High gear 100

    Low gear 200

Traverse for one turn of handwheel:

    High gear .... 3.6 deg (63.8 mils)

    Low gear .... 1.8 deg (31.9 mils)

Effort required at elevating handwheel (in.-lb):

To Elevate	High Gear	Low Gear
0 deg	55	110
20 deg	110	160
40 deg	192	110
60 deg	214	55
80 deg	209	50

To Depress	High Gear	Low Gear
0 deg	275	220
20 deg	193	28
40 deg	138	50
60 deg	110	77
80 deg	165	77

INTRODUCTION

Effort required at traversing handwheel (in.-lb):

To Traverse Left	High Gear	Low Gear
0 deg	55	39
90 deg	28	6
180 deg	11	11
270 deg	22	17

To Traverse Right	High Gear	Low Gear
0 deg	10	6
90 deg	44	17
180 deg	61	44
270 deg	20	17

Time to elevate from minus 3 to plus 85 degrees:
High gear .... 15.02 sec
Low gear .... 25.90 sec

Time to depress from plus 85 to minus 3 degrees:
High gear .... 21.44 sec
Low gear .... 34.90 sec

Time to traverse 360 degrees:
High gear .... 33.90 sec
Low gear .... 69.79 sec

Over-all dimensions in firing position:
Length .... 19 ft
Height .... 6.9 ft
Width .... 16.87 ft w/outriggers

Over-all dimensions in traveling position:
Length .... 25.5 ft w/drawbar
Height .... 7.9 ft
Width (front) .... 7.20 ft
Width (rear) .... 7.60 ft

Length of outriggers .... 4.8 ft
Number of bogies 2
Type of bogies .... Single axle. Single wheels on
front; dual wheels on rear
Weight of front bogie.. 1,825 lb
Weight of rear bogie .... 2,645 lb
Pneumatic tire size .... 32 in. x 6½ in. (6½ extra 20);
also marked 7:50 x 20
Wheel base .... 13.75 ft
Type of brakes .... Vacuum air brakes on all wheels;
hand-operated parking brakes
on rear wheels also

GERMAN 88-MM ANTIAIRCRAFT GUN MATERIEL

Type and number of jacks .... 4 jacks integral with mount for leveling bottom carriage; one on each end of outriggers and carriage

Leveling .... 4.5 deg leveling either side of horizontal

Road clearance .... 1.14 ft

Tread (front) .... 5.8 ft

Tread (rear) .... 6 ft

Height of axis of bore above ground (firing) .... 5 ft

Time to change from traveling to firing position....2½ min with 6-man crew (approx.)

Time to change from firing to traveling position....3½ min with 6-man crew (approx.)

Weight of entire carriage .... 16,325 lb

Rear wheel reactions .... 9,830 lb

Front wheel reactions .... 6,510 lb

Type of equilibrators ....Spring type with built-in spring compressors

d. Essential Translations.

Schnell .... Quick

Normal .... Normal

Automatik .... Automatic

Hand .... Hand

Wiederspannen .... Recock

Los .... Loose

Fest .... Tight

Linksgewinde .... Left-hand thread

Mundung .... Muzzle

Feuer .... Fire

Sicher .... Safe

CHAPTER 2

GERMAN 88-MM ANTIAIRCRAFT GUN AND MOUNT

Section I

DESCRIPTION AND FUNCTIONING OF GUN

4. GERMAN 88-MM ANTIAIRCRAFT GUN.

a. The German 88-mm antiaircraft gun consists of a detachable breech ring with a half-length outer tube, a half-length inner lock tube, and a loose three-piece liner.

b. The liner separates into three sections, one division being two-thirds of the rifled length back from the muzzle, and the other division being approximately 6 inches to the rear of the origin of rifling. Instead of replacing the entire length of liner as is the practice in this country, economy is achieved by replacing just that section of the liner which receives the most wear, i.e., the forcing cone section.

c. The front and center sections of the liner are keyed in place so as to aline the rifling and prevent relative rotation. This joint does not have any seal other than that provided by close tolerance machining. The center and rear sections are merely overlapped and not keyed in place as there is no rifling to aline (fig. 3).

d. The three sections are alined end to end and then fitted into the inner tube (fig. 4). This tube serves to prevent lateral move-

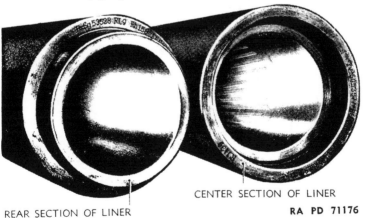

REAR SECTION OF LINER

CENTER SECTION OF LINER

RA PD 71176

Figure 3 — Center and Rear Sections of Liner

GERMAN 88-MM ANTIAIRCRAFT GUN MATERIEL

RA PD 71177

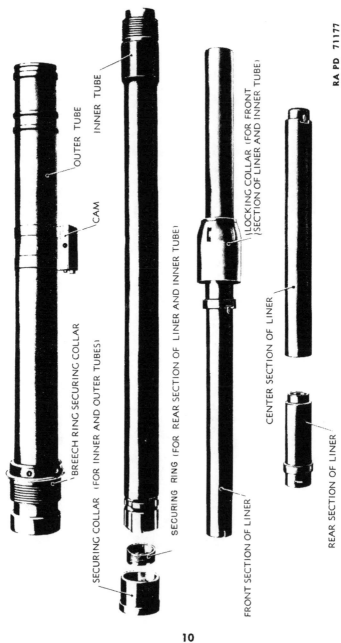

OUTER TUBE

INNER TUBE

CAM

BREECH RING SECURING COLLAR

SECURING COLLAR (FOR INNER AND OUTER TUBES)

SECURING RING (FOR REAR SECTION OF LINER AND INNER TUBE)

LOCKING COLLAR (FOR FRONT SECTION OF LINER AND INNER TUBE)

CENTER SECTION OF LINER

FRONT SECTION OF LINER

REAR SECTION OF LINER

Figure 4 — Tube and Liners

DESCRIPTION AND FUNCTIONING OF GUN

RA PD 71178

Figure 5 — Method of Securing Chamber Sections of Liner to Inner Tube

ment and to prevent rotation between the rear of chamber sections and other sections of the liner. The locking collar (fig. 4) prevents forward movement, and the locking ring (fig. 4) prevents movement to the rear. See figure 5 for method of fastening the chamber sections of liner to the inner tube. When the locking ring and collar are fully tightened, the liner sections are drawn up snugly and the joints offer little or no resistance to the passage of the projectile. The female threads in the locking collar are left-hand as indicated by the word "LINKSGEWINDE" (fig. 6). The collar is rotated in the direction of "LOOSE" ("LOS") for removing and in the direction of "TIGHT" ("FEST") for tightening.

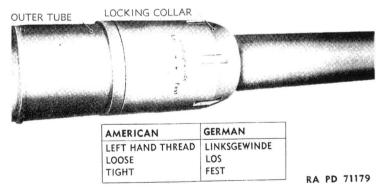

AMERICAN	GERMAN
LEFT HAND THREAD	LINKSGEWINDE
LOOSE	LOS
TIGHT	FEST

RA PD 71179

Figure 6 — Markings on Locking Collar

GERMAN 88-MM ANTIAIRCRAFT GUN MATERIEL

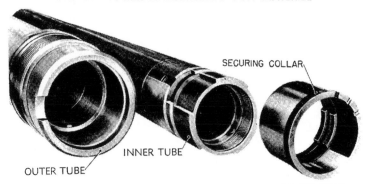

RA PD 71180

Figure 7 — Method of Securing Inner and Outer Tubes

e. The inner tube which contains the liner sections is slipped into the outer tube (fig. 4). The latter tube has fastened to it the forward end of the slides. The breech ring also fits on the outer tube. The inner tube is secured by the locking collar (fig. 4) to prevent forward and rearward motion. See figure 7 for the method of securing the inner tube to the outer tube.

f. The breech ring does not screw on the tube as is the practice in this country. Instead, the breech ring slides over the tube until it is seated and then the securing collar (fig. 4) draws it up tightly. This eliminates the need for rotating the tube or breech ring. In order to prevent rotation of the outer tube and the two locking collars, keys are provided. See figure 8 for installation of the keys.

5. BREECH MECHANISM.

a. The breech mechanism is of the horizontal sliding breechblock type actuated by a breech operating spring permitting semiautomatic or manual operation (fig. 9). The breechblock slides in a rectangular breech ring which is bored to receive the outer tube and the breechblock. Channels are machined into the bottom of the ring to permit installation of recoil slide pads. The recoil piston rod lug is made an integral part of the breech ring.

b. With the breech mechanism set for semiautomatic operation, a round of ammunition, when pushed in the breech recess of the gun, will trip the extractors and allow the breechblock to close under the action of the breech actuating spring. When the gun is fired, and recoils, the breechblock actuating shaft, which is operated by the breech operating crank, is rotated by the cam on the side of the cradle (fig. 10). This action winds up the lower breech opening spring and draws the intermediate plate away from its stop.

12

DESCRIPTION AND FUNCTIONING OF GUN

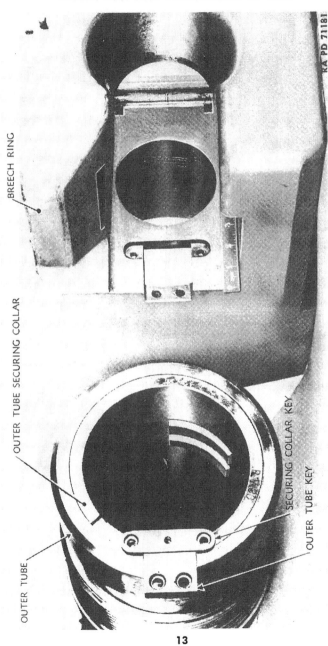

RA PD 71181

BREECH RING

OUTER TUBE SECURING COLLAR

OUTER TUBE

SECURING COLLAR KEY

OUTER TUBE KEY

Figure 8 — Method of Fastening Breech Ring to Outer Tube and Securing Collar

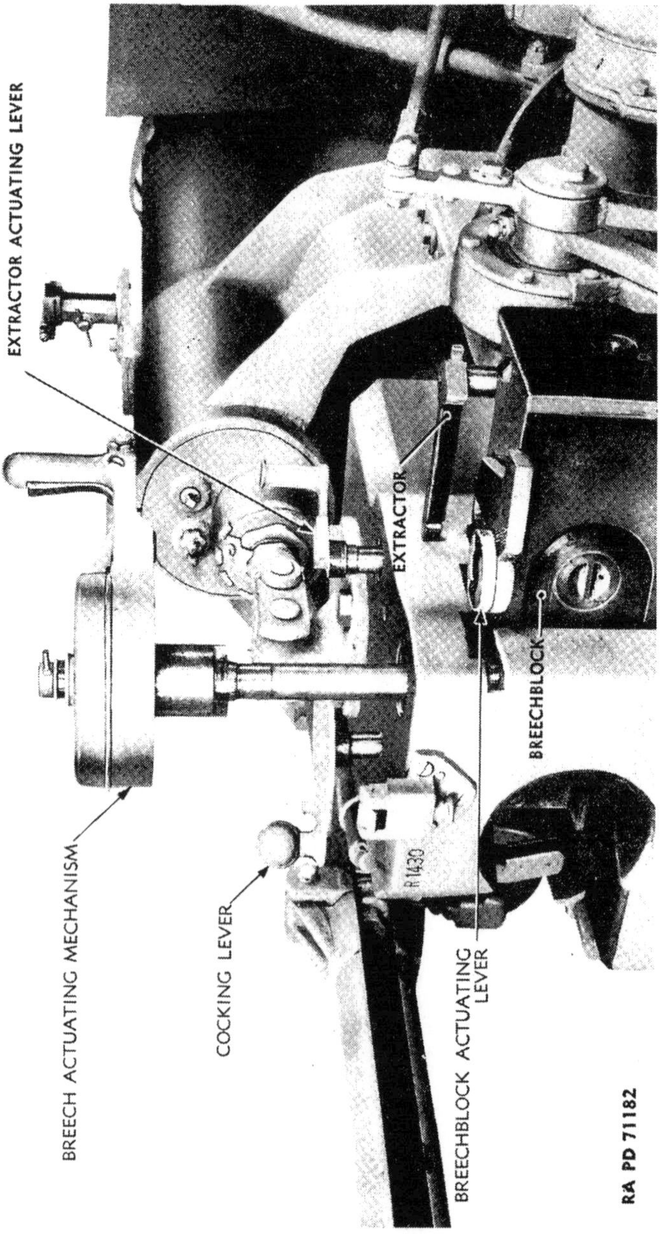

GERMAN 88-MM ANTIAIRCRAFT GUN MATERIEL

EXTRACTOR ACTUATING LEVER

EXTRACTOR

BREECHBLOCK

BREECH ACTUATING MECHANISM

COCKING LEVER

BREECHBLOCK ACTUATING LEVER

RA PD 71182

Figure 9 — Breech Mechanism

DESCRIPTION AND FUNCTIONING OF GUN

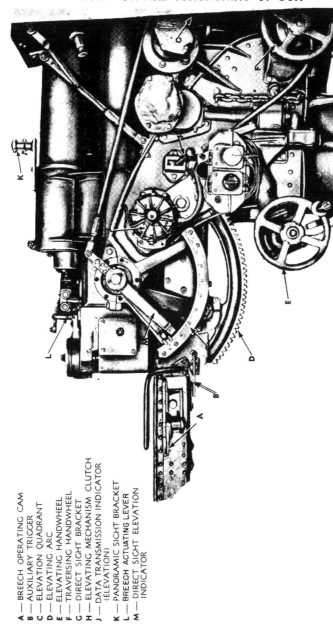

RA PD 71183

Figure 10 — Breech End — Right Side

A — BREECH OPERATING CAM
B — AUXILIARY TRIGGER
C — ELEVATION QUADRANT
D — ELEVATION ARC
E — ELEVATING HANDWHEEL
F — TRAVERSING HANDWHEEL
G — DIRECT SIGHT BRACKET
H — ELEVATING MECHANISM CLUTCH
J — DATA TRANSMISSION INDICATOR
     (ELEVATION)
K — PANORAMIC SIGHT BRACKET
L — BREECH ACTUATING LEVER
M — DIRECT SIGHT ELEVATION
     INDICATOR

c. When the breech operating crank is in the straight position of the cam path, the catch on the top spring cover disengages the lug retaining the breech actuating mechanism closed. Then the actuating shaft is free to rotate to the open position under the action of the breech opening spring, taking the crank with it and so opening the breechblock by the action of the breechblock actuating lever. The extractors are tripped and the empty case is ejected.

d. In the full open position, the compression of the breech opening spring is taken between the lug on the intermediate plate and its mating step. Any further opening motion of the breechblock is then taken up by the breech closing spring which, at this stage, acts as a breechblock buffer stop.

e. The breechblock is held open by the action of the extractors hooking on the recesses in the breechblock against the action of the upper spring which has been further wound up during recoil.

f. With the breech mechanism set for hand operation, the springs are disengaged from the breech actuating mechanism and the breechblock may then be opened or closed with no spring influence.

g. The breechblock may be closed without loading a round by the action of the extractor actuating shaft. This is a splined shaft extending through both extractors. The extractors are tripped and removed from the recesses in the breechblock by rotating the shaft by hand (fig. 11).

6. FIRING MECHANISM.

a. The firing mechanism is composed of the percussion mechanism, percussion mechanism release assembly, the cocking lever assembly, and the cradle firing mechanism.

b. The percussion mechanism is composed of the firing spring retainer, firing spring, firing pin, and firing pin holder (fig. 12). This group is held in the axial hole of the breechblock by a lug on the firing spring retainer engaging a mating groove in the breechblock. The percussion mechanism is operated through the percussion mechanism release assembly.

c. The percussion mechanism release assembly is located in various recesses of the breechblock. This assembly is composed of the cocking arm, operating rod, operating rod spring, safety stop lever, operating rod guide, sear, sear spring, and sear operating lever. Through this assembly, the percussion mechanism is cocked either automatically or normally.

d. Cocking of the percussion mechanism automatically is accomplished during opening of the breech, with the cocking lever in the "FIRE" ("FEUER") position. As the breechblock slides to the right in recoil or hand operation, the cocking arm is engaged by the breechblock actuating lever. As the actuating lever rotates on the actuating shaft, the cocking arm is also rotated. The cocking lug on the firing

DESCRIPTION AND FUNCTIONING OF GUN

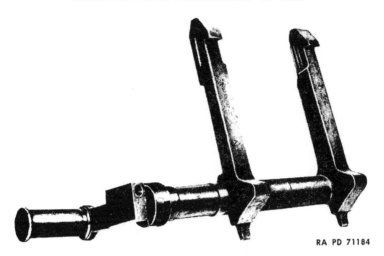

RA PD 71184

Figure 11 — Extractors and Actuating Shaft

pin holder is engaged and slid to the rear, compressing the firing spring. When the holder reaches the cocked position, the sear lug on the holder is engaged by the notch of the sear, which holds the mechanism cocked.

e. Manual cocking of the percussion mechanism is accomplished by the cocking lever (fig. 13). This lever is located on top of the breech ring. The breech must be closed when manual cocking is performed. The cocking lever serves the same purpose as the actuating lever, i.e., to rotate the cocking arm. However, in this case, if the cocking lever is kept in the rear, the firing mechanism will not operate but will be on "SAFE" ("SICHER") (fig. 13) because the lug on the cocking arm will not have cleared the lug on the firing pin holder.

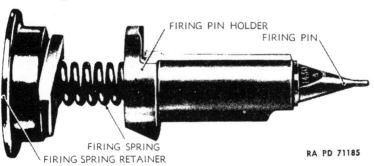

FIRING PIN HOLDER

FIRING PIN

FIRING SPRING

FIRING SPRING RETAINER

RA PD 71185

Figure 12 — Percussion Mechanism

17

GERMAN 88-MM ANTIAIRCRAFT GUN MATERIEL

Figure 13 — Breech Mechanism — Top View

Thus the cocking lever also serves as a safety. The arc described by the cocking lever during manual cocking of the percussion mechanism is marked "WIEDERSPANNEN" which, freely translated, means "recock."

DESCRIPTION AND FUNCTIONING OF GUN

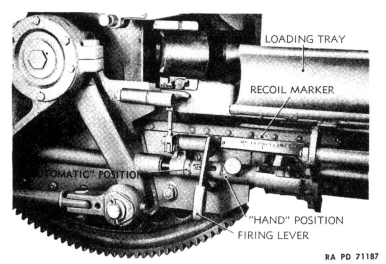

LOADING TRAY

RECOIL MARKER

"AUTOMATIC" POSITION

"HAND" POSITION

FIRING LEVER

RA PD 71187

Figure 14 — Loading Tray Interlock Mechanism

f. The cradle firing mechanism is located on the left side of the cradle. Raising the firing lever (fig. 14) of the cradle firing mechanism forces the lug up at the end of the sear operating lever. This, in turn, pushes the sear down against the sear spring disengaging the sear lug. The firing pin holder and firing pin, thus released, are driven forward by the compressed firing spring to fire the primer in the cartridge.

g. The firing mechanism will not operate unless the breechblock is fully closed. If the breechblock is not fully closed, the operating rod will not be fully in position against the compression of the operating spring rod. This will prevent the safety stop lever from rotating, and hence will not permit clearance for the firing pin holder to move forward, thus rendering the firing mechanism inoperative.

GERMAN 88-MM ANTIAIRCRAFT GUN MATERIEL

CHAPTER 2

GERMAN 88-MM ANTIAIRCRAFT GUN AND MOUNT (Cont'd)

Section II

DESCRIPTION AND FUNCTIONING OF RECOIL MECHANISM

Paragraph

7. DESCRIPTION AND FUNCTIONING OF RECOIL MECHANISM.

a. General. The recoil mechanism is an independent hydropneumatic system. The recuperator cylinder, which is entirely separate from the recoil cylinder, is filled and charged with gas and liquid in direct contact. The recoil cylinder is of the control rod type with a secondary control rod regulating recoil length. Both recuperator and recoil cylinders are supported by the cradle, and the pistons are connected to the top and bottom of the breech ring, respectively.

RA PD 71188

Figure 15 — Recuperator Cylinder

b. Recuperator Cylinder.

(1) The recuperator cylinder (fig. 15) is secured to the cradle above the piece. A liquid cylinder is fitted eccentrically in the bottom of the outer gas cylinder. The center lines of both cylinders are parallel. The liquid cylinder is completely filled with a glycerine-water solution, and the rest of the mechanism is charged with nitrogen to the proper pressure.

DESCRIPTION AND FUNCTIONING OF RECOIL MECHANISM

RA PD 71189

Figure 16 — Recoil Cylinder

(2)   Upon recoil, the recuperator cylinder rod and piston are brought to the rear by the recoiling gun; and the liquid is transferred, by the piston, from the liquid cylinder into the gas cylinder. The gas is compressed by the decreased volume in the cylinder, thus opposing the energy of recoil. While the recuperator cylinder controls a portion of the recoiling energy, the recoil cylinder controls the remainder of the recoiling energy in addition to controlling the length of recoil. In counterrecoil, the motivating force is the expanding gas tending to force the liquid back into the liquid cylinder, thus activating the recuperator cylinder piston. The force of counterrecoil is dampened by the recoil cylinder. After several rounds have been fired, the gas and liquid have emulsified. This condition, however, does not alter the volume pressure relationship, and the liquid is still effective for its original purpose of supplying an adequate pressure seal. The ports in the end of the liquid cylinder are not throttling orifices, and the state of emulsification has no effect on the recoil action.

(3)   The piston rod is hollow to eliminate the vacuum which would be caused by the sealed cylinder and plug. This hollow opening also permits exit of the atmospheric air in back of the piston head. The washers are of U-shaped leather and use U-shaped brass spacers. The whole is secured by a large lock nut.

c.   Recoil Cylinder.

(1)   The recoil cylinder (fig. 16) is located beneath the gun inside the cradle. The cylinder is filled with liquid at atmospheric pressure. The cylinder and the control rod remain stationary. In recoil, the piston rod and counterrecoil control rod move with the breech ring. As the weapon recoils, part of the fluid is forced through the orifices in the piston head and through the control grooves in the recoil control rod. Another portion of the fluid passes through the valve in the control bushing and fills the increasing hollow space behind the head of the recoil control rod. The pressure of the liquid through the constantly narrowing grooves takes up most of the force of recoil and gradually brings the gun to a standstill. Part of the force of recoil is also taken up in the increase of air pressure in the recuperator cylinder.

GERMAN 88-MM ANTIAIRCRAFT GUN MATERIEL

RA PD 71190

Figure 17 — Recoil Control Linkage

(2) The counterrecoil action is activated by the expanding air in the recuperator cylinder. The braking liquid which is now in front of the recoil piston head runs back through the control bushing of the recoil control rod. The piston rod slides back over the recoil control rod and the counterrecoil control rod penetrates deeper into the recoil control rod, displacing the liquid in the latter. The valve being closed, the fluid is forced through the grooves in the counterrecoil control rod and the holes in the head. The force of counterrecoil is thus reduced, and the gun comes to rest without shock.

(3) To change the length of recoil as required by high angle fire, the recoil control rod is rotated by the length of recoil control linkage (fig. 17). The linkage is operated when the cradle is elevated and serves to rotate the throttling grooves, thus varying the port area over the whole length of recoil.

CHAPTER 2

GERMAN 88-MM ANTIAIRCRAFT GUN
AND MOUNT (Cont'd)

Section III

DESCRIPTION AND FUNCTIONING OF MOUNT

8. GENERAL.

a. The German 88-antiaircraft gun mount is a mobile unit carried in traveling position by two bogies (fig. 18). This gun is a dual-purpose weapon. It can be fired from the bogie wheels as an antitank weapon, or the bogies can be removed to emplace the weapon for antiaircraft fire (figs. 19 and 20). The mount consists mainly of the bottom carriage, side outriggers, leveling jacks, top carriage leveling mechanism, pedestal, top carriage, cradle, equilibrators, traversing mechanism, elevating mechanism, bogies, and rammer.

9. BOTTOM CARRIAGE.

a. The bottom carriage is of box-section type construction, welded, and riveted. The bottom carriage is designed to form a chassis for connection to the bogies in traveling. For stability during firing, a large base area is incorporated into the design of the bottom carriage, with front and rear outriggers being integral (fig. 21). Greater stability is obtained by hinging side outriggers to the bottom carriage. The interior of the bottom carriage provides space for storing tools and accessories and for housing the electrical wiring.

b. The pedestal is bolted to the enlarged central portion of the bottom carriage. This portion also houses the handwheels used to level the top carriage. The data transmission junction box is located at the rear end of the bottom carriage (fig. 22). The gun muzzle rest for road transportation is supported at the front end. Two lugs at each end of the bottom carriage are provided to suspend the mount from the bogies. Hooks at each end of the bottom carriage are provided to engage the bogie chains (fig. 21).

GERMAN 88-MM ANTIAIRCRAFT GUN MATERIEL

RA PD 71191

FRONT BOGIE

REAR BOGIE

Figure 18 — German 88-mm Antiaircraft Gun — Right Side — Traveling Position

DESCRIPTION AND FUNCTIONING OF MOUNT

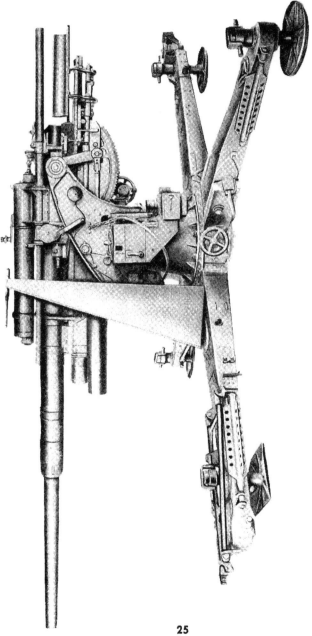

RA PD 71192

Figure 19 — German 88-mm Antiaircraft Gun — Zero-degree Elevation

GERMAN 88-MM ANTIAIRCRAFT GUN MATERIEL

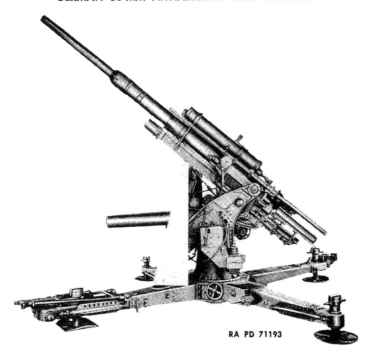

RA PD 71193

Figure 20 — German 88-mm Antiaircraft Gun — Left Side View

10. SIDE OUTRIGGERS.

a.   The side outriggers are of the same construction as the bottom carriage, i.e., welded and riveted. They hinge to the bottom carriage and provide stability when firing in traverse other than directly to the rear or front. In traveling position, they are swung to a vertical position and secured against the mount. In firing position, the side outriggers are let down and secured in position by half-round locking pins (fig. 23). The side outriggers are provided with leveling jacks and stakes at the extremities as are the outriggers of the bottom carriage (fig. 24).

11. LEVELING JACKS.

a.   The leveling jacks (fig. 25) are of a simple lead screw construction. Four leveling jacks are provided, one at the extremity of each of the side outriggers (fig. 24) and of each of the bottom carriage outriggers (fig. 21). They serve to distribute firing loads evenly when the mount is on uneven ground.

DESCRIPTION AND FUNCTIONING OF MOUNT

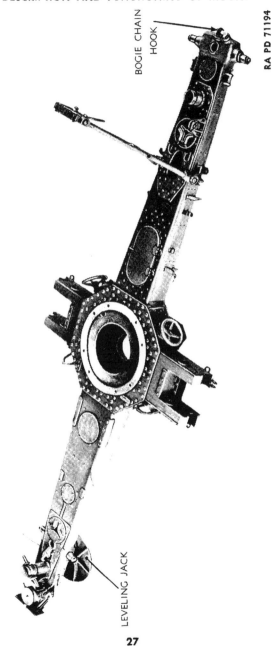

BOGIE CHAIN HOOK

RA PD 71194

LEVELING JACK

Figure 21 — Bottom Carriage

27

GERMAN 88-MM ANTIAIRCRAFT GUN MATERIEL

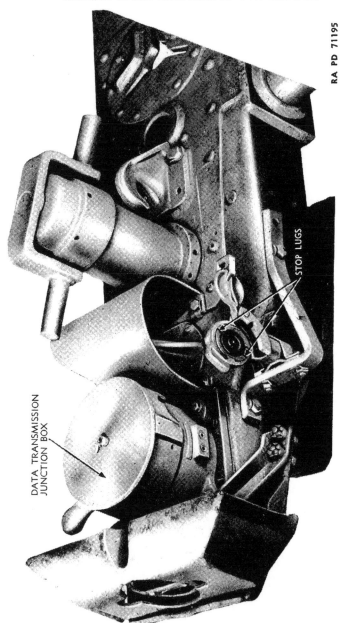

RA PD 71195

Figure 22 — Rear End of Bottom Carriage

DESCRIPTION AND FUNCTIONING OF MOUNT

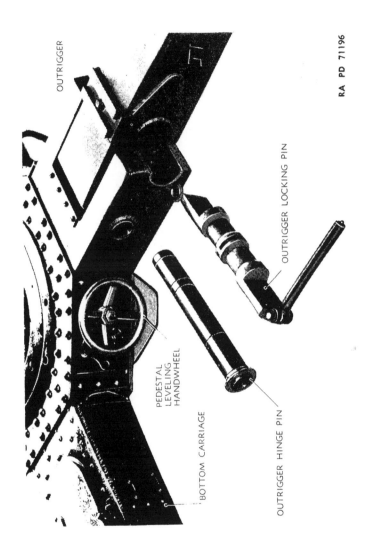

OUTRIGGER

RA PD 71196

OUTRIGGER LOCKING PIN

PEDESTAL
LEVELING
HANDWHEEL

BOTTOM CARRIAGE

OUTRIGGER HINGE PIN

Figure 23 — Bottom Carriage Outrigger and Connecting Pins

GERMAN 88-MM ANTIAIRCRAFT GUN MATERIEL

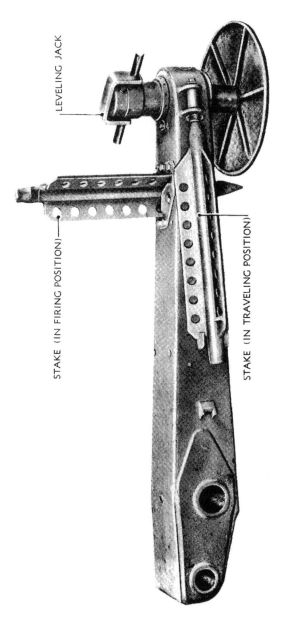

RA PD 71197

LEVELING JACK

STAKE (IN FIRING POSITION)

STAKE (IN TRAVELING POSITION)

Figure 24 — Outrigger, Showing Position of Stakes in Firing and Traveling Position

DESCRIPTION AND FUNCTIONING OF MOUNT

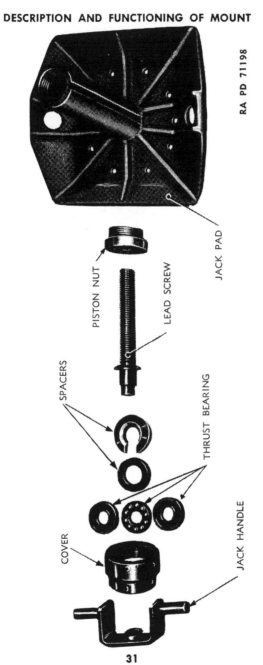

RA PD 71198

PISTON NUT

LEAD SCREW

JACK PAD

SPACERS

THRUST BEARING

COVER

JACK HANDLE

Figure 25 — Details of Firing Jack

GERMAN 88-MM ANTIAIRCRAFT GUN MATERIEL

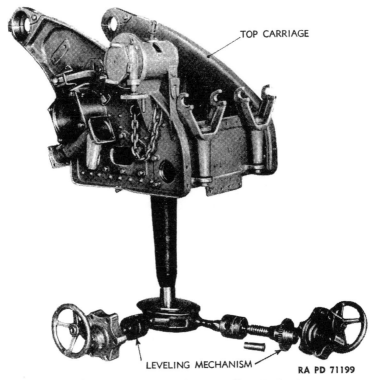

TOP CARRIAGE

LEVELING MECHANISM

RA PD 71199

Figure 26 — Top Carriage Leveling Mechanism

12. TOP CARRIAGE LEVELING MECHANISM.

a. The top carriage leveling mechanism (fig. 26) is operated by handwheels in the enlarged central portion of the bottom carriage. The mechanism operates the linkages that tip the top carriage about the two centers of rotation, thereby alining the gun trunnions at a horizontal position. A level indicator is provided on the pedestal (fig. 39).

13. PEDESTAL.

a. The pedestal (fig. 27) is made in three sections, namely, the pedestal, leveling universal, and traversing ring. The pedestal is of welded construction. The leveling universal is suspended in the pedestal trunnion bearings by trunnions and is tipped about the trunnions and secondary pivots. The traversing ring is bolted directly to the top of the leveling universal. The pedestal is bolted to the bottom carriage and supports the top carriage.

DESCRIPTION AND FUNCTIONING OF MOUNT
LEVELING UNIVERSAL

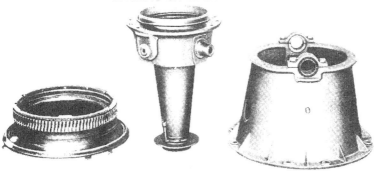

TRAVERSING RING   PEDESTAL   RA PD 71200

Figure 27 — Components of Pedestal Assembly

b. An adjustable azimuth scale is provided for the orientation of the weapon. The leveling universal houses the self-alining roller bearing gimbal and ball thrust bearing on the pintle of the top carriage (fig. 28).

14. TOP CARRIAGE.

a. The top carriage (fig. 26) is of welded construction. The forged hollow pintle is welded to the top carriage and houses the data trans-

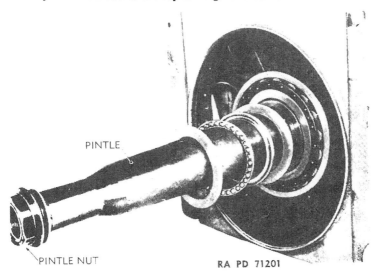

PINTLE

PINTLE NUT   RA PD 71201

Figure 28 — Pintle and Bearing Arrangement

33

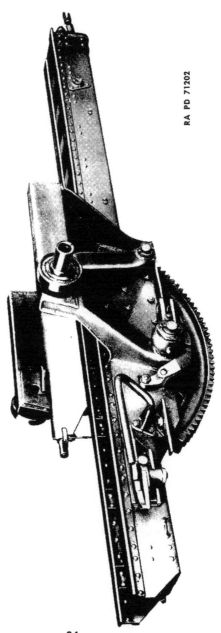

RA PD 71202

Figure 29 — Cradle

DESCRIPTION AND FUNCTIONING OF MOUNT

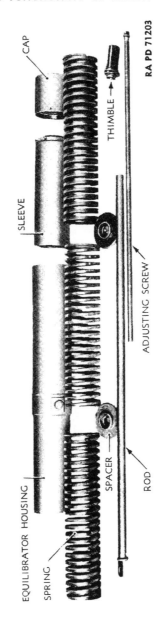

RA PD 71203

Figure 30 — Details of Equilibrator

GERMAN 88-MM ANTIAIRCRAFT GUN MATERIEL

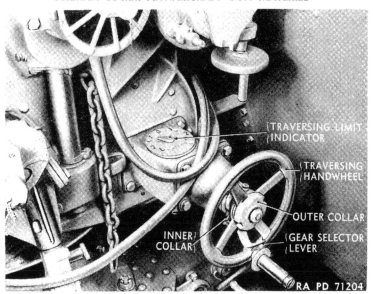

Figure 31 — Traversing Limit Indicator

Figure 32 — Elevating Mechanism Clutch Disassembled

DESCRIPTION AND FUNCTIONING OF MOUNT

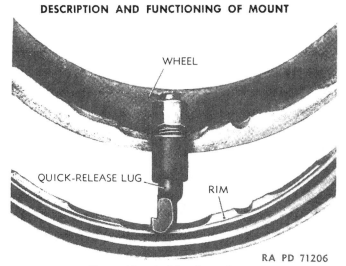

Figure 33 — Quick-release Lug

mission cable. The top carriage rests directly on the leveling universal of the pedestal and is kept in place by the gimbal bearing. The nut at the end of the pintle (fig. 28) prevents any vertical motion.

b. The cradle trunnion housings, azimuth and elevation mechanism housings, direct sight elevating housing, and equilibrator

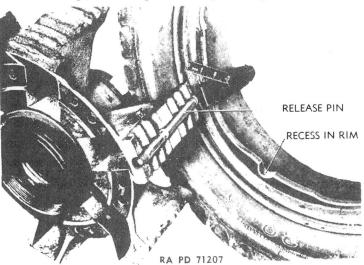

Figure 34 — Quick-release Device

GERMAN 88-MM ANTIAIRCRAFT GUN MATERIEL

trunnion supports are all welded to the top carriage. The leveling mechanism is fastened to the roller bearing at the lower end of the pintle.

15. CRADLE.

a. The cradle (fig. 29) is of rectangular-trough type section, welded, and riveted. The slides of the cradle support and guide the gun during recoil and counterrecoil. Trunnions are welded directly to the side frames and support the self-alining roller bearings. The equilibrator rod is fastened to the rear and below the trunnions by means of two clevis joined by a pin running through the cradle. The single elevating arc is fixed beneath the cradle by means of the equilibrator clevis pin and another pin just forward. Thus the arc is fastened to the cradle at two points and is readily replaced by removing the two pins. The breech operating cam and the auxiliary trigger mechanism are both fixed to the rear right side of the cradle. The loading tray is fastened to the rear left side as are the firing lever and recoil marker (fig. 14).

16. EQUILIBRATORS.

a. Two spring equilibrators are used to balance the muzzle preponderance of the gun. The equilibrators are suspended from the top carriage by trunnions and are fastened to the cradle by a clevis. In each unit there are three rectangular cross-section wire springs separated by spacers. The equilibrator rod also serves as a spring compressor and adjusting screw. Each unit is encased in a telescoping housing (fig. 30).

17. TRAVERSING MECHANISM.

a. The traversing handwheel is located on the right side of the mount (fig. 10). The traversing mechanism may be operated in either high or low speed. For changing from one speed to the other, a gear selector lever is provided at the handwheel (fig. 31).

b. A 360-degree traverse is permitted by the traversing ring. An indicator, located above the traversing handwheel, shows when the mount has made up to two complete revolutions in either direction (fig. 31). A Belleville spring stop at the left side of the top carriage prevents rotation in excess of two complete turns in any one direction. This is to prevent tangling of the data transmission cable. The azimuth data transmission indicator is geared directly to the circular rack just above the traversing ring.

18. ELEVATING MECHANISM.

a. The elevating handwheel is located on the right side of the carriage (fig. 10). Motion is transmitted from the handwheel, through gears, to the elevating pinion which engages the elevating rack, thereby elevating or depressing the gun. The elevating mechanism may be operated in either high or low speed. For changing from one speed

DESCRIPTION AND FUNCTIONING OF MOUNT

to the other, a gear selector lever is provided at the handwheel similar to the one on the traversing handwheel.

b. A clutch (fig. 32) is provided as a means of disengaging the elevating mechanism (fig. 10) from the elevating arc to prevent transmission of road shocks to the elevating gear system during traveling. The clutch mechanism is designed to prevent excessive wear on the edges of the clutch teeth when improperly meshed. The operational design of the clutch prevents meshing until the teeth are correctly alined. The clutch alining gear is always in contact with the spur gear as the clutch fork moves the clutch body on; the central pin moves forward at the same time. The clutch teeth will not engage until this pin enters the receiving hole in the clutch alining gear. This hole is properly concentric in only one position. At this position the pin will properly seat and the clutch will engage. When both sides of the clutch engage, there is no relative rotation between the alining gear and the clutch, and the holes remain alined until the clutch is disengaged for traveling position.

19. BOGIES.

a. The front and rear bogies are of welded construction, single axle type. The front bogie is fitted with 7-leaf transverse spring and has single wheels. The rear bogie is fitted with conventional 11-leaf suspension springs and has dual wheels.

b. The wheels are of cast spoke construction with twin detachable rims on the rear bogie and single detachable rims on the front bogie. The wheel spoke casting is fastened to the brake drum. The brake shoes are castings with the lining riveted to the outer surface. The shoes are actuated by a cam as is the practice in this country. The rims are fastened to the wheels by means of quick-release lugs (fig. 33). The lugs are loosened and then moved along the rim to bring them in line with recesses provided for the purpose and then removed. The wheels are removed from the spokes by another quick-release device (fig. 34). The release pin is pulled out until the recessed shoulder permits a quarter turn of the wheel and then the wheel may be removed by pulling straight off.

c. The mount is equipped with air brakes on all wheels. The rear bogie is provided with a seat from which the hand brake lever may be operated in case of emergencies. The stop lugs (fig. 22) on the German air hose connections must be filled slightly to allow the ones on American prime movers to be inserted. At best, only a loose connection is possible, thus resulting in a leakage of air.

d. The adjustable height drawbar is fastened to the front bogie axle projection and also controls the action of the radius bars (fig. 46). The lunette on the drawbar is large enough to fit the pintle on American prime movers. The bogies are equipped to take a single tube transporter bar to connect the two bogies when removed from the mount so as to make an improvised trailer.

GERMAN 88-MM ANTIAIRCRAFT GUN MATERIEL

Figure 35 — Rammer Mechanism

20. RAMMER.

a. To facilitate the loading of rounds at high angles of elevation, an automatic rammer is provided (fig. 35). This rammer is mounted on the left top of the cradle and is actuated by a hydropneumatic cylinder. The rammer head is cocked automatically during counter-recoil and is released by the action of the hand-operated loading tray.

b. The actuating mechanism (fig. 36) utilizes gas and liquid in direct contact as in the recuperator cylinder. In this instance the cylinder is movable and the piston is fixed to the cradle. The cylinder has a removable inner liner, eccentrically located, but with its axis parallel with the axis of the outer cylinder. The system is filled with liquid to the level of the top of the inner cylinder and with nitrogen under pressure. A gas check valve is used for buffer action.

c. As the gun returns to battery in counterrecoil, a cam on the outer tube (fig. 4) engages a catch on the cylinder. The force of counterrecoil forces the cylinder forward until the gun returns to battery. At this point the cam on the tube is disengaged and the loading tray interlock prevents the cylinder from returning the position. The motivating force for ramming is now the mixture of liquid and gas under pressure.

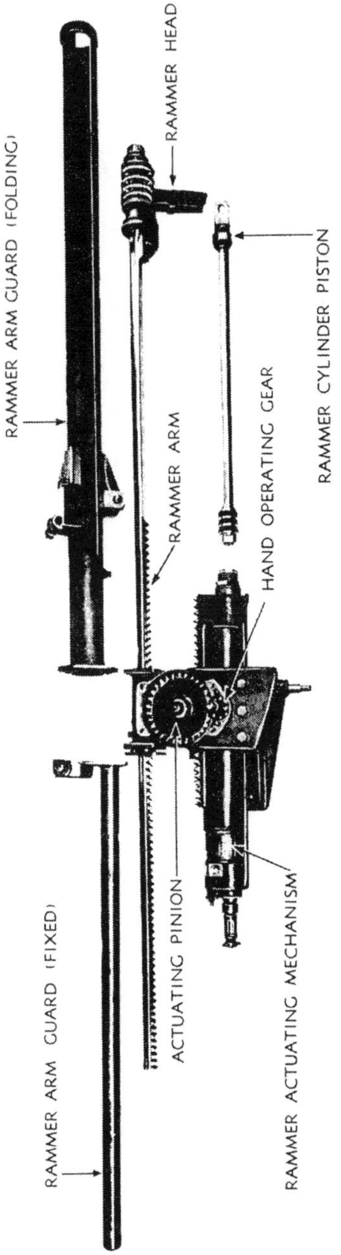

RA PD 71209

RAMMER ARM GUARD (FOLDING)

RAMMER HEAD

RAMMER ARM

HAND OPERATING GEAR

RAMMER CYLINDER PISTON

ACTUATING PINION

RAMMER ACTUATING MECHANISM

RAMMER ARM GUARD (FIXED)

Figure 36 — Automatic Rammer Assembly

GERMAN 88-MM ANTIAIRCRAFT GUN MATERIEL

d. As the cylinder moves forward with the gun in counterrecoil, the rack and pinion linkage (fig. 36) actuates the ramming arm in the opposite direction. Thus, when the gun is in battery and the cylinder fully cocked, the ramming arm is fully extended and the mechanism is in ramming position. The loading tray is hand-operated and is mounted on two supporting lugs on the left side of the cradle. When the gun is fired, the loading tray is outboard of the cradle slides, and the new round may be placed on the tray at any time. As the gun returns to battery and the rammer head is fully cocked, the round is placed in loading position by grasping the handles on the side of the tray and pushing the tray over on its pivot until the axis of the round is on the same line as the axis of the bore. At this point the loading tray interlock (fig. 14) is released and the expanding gas forces the rammer cylinder back along the piston; thus the rammer arm is rapidly withdrawn seating the round. The loading tray interlock will not permit the trigger mechanism to operate until the loading tray is returned to the outboard position. There is an "AUTO-MATIC" ("AUTOMATIK") position (fig. 14) on the interlock that will permit the trigger mechanism to function automatically when the loading tray is returned to the loading position. When the interlock is on "HAND," the trigger handle must be operated manually in order to fire the piece.

e. The rammer head (fig. 36) is permitted to swivel on the rammer arm. Thus, in order to ram the round, the rammer head is returned by hand to a position in which the base of the round may be engaged. When the round is rammed, the horizontal sliding breechblock strikes the end of the head and throws it over into a position that will enable the recoiling gun to clear.

f. The rammer arm is protected in all positions by a folding guard which also serves as a guide. When the gun has been fired for the last time before preparing for a change of location, the loading tray interlock may be operated without a round in the tray, thus releasing the pressure on the rammer and permitting the end of the guard to be folded back. In order to cock the rammer before the first round is fired, a removable handle (fig. 35) is available to rotate the rack and pinion linkage and thus force the cylinder back until the loading tray interlock will take effect. From this point on, all operations are the same as previously noted.

g. At elevations above 45 degrees the air buffer operation is reduced to obtain additional energy for ramming by permitting the air to escape at a faster rate. This is accomplished by setting the buffer valve, at the front of the rammer cylinder (fig. 35), to "FAST" ("SCHNELL"). For elevations below 45 degrees, the valve should be set to "NORMAL" ("NORMAL").

CHAPTER 2

GERMAN 88-MM ANTIAIRCRAFT GUN AND MOUNT (Cont'd)

Section IV

OPERATION

21. TO PLACE THE WEAPON IN FIRING POSITION.

a. The piece may be fired from the wheels but must be emplaced for high angle fire. To fire from the wheels:

(1) Unlimber the prime mover from the drawbar.

(2) Set the hand brakes on the rear bogie.

RA PD 71210

Figure 37 — Engaging Elevation Gear Clutch

GERMAN 88-MM ANTIAIRCRAFT GUN MATERIEL

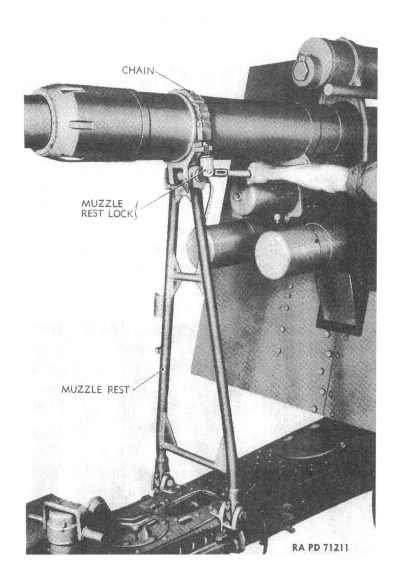

Figure 38 — Releasing Muzzle Rest

44

OPERATION

LEVEL
INDICATOR

CROSS-LEVELING
HANDWHEEL

RA PD 71212

Figure 39 — Leveling Top Carriage

45

GERMAN 88-MM ANTIAIRCRAFT GUN MATERIEL

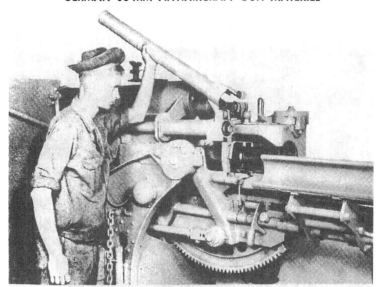

Figure 40 — Unfolding Rammer Guard   RA PD 71213

(3) Engage the elevating gear clutch by pulling the clutch lever to its downward position (fig. 37).

(4) Release the muzzle rest (fig. 38) by:

(a) Unscrewing the muzzle rest lock so that the chain may be swung over the barrel.

(b) Elevating the gun slightly so that the muzzle rest may be pushed forward and down onto the bottom carriage.

(5) Level the top carriage by cross-leveling handwheels, using the level indicator for reference (fig. 39).

(6) Unfold the rammer guard (fig. 40) to the operating position (fig. 41).

(7) Cock the rammer assembly by rotating the rammer crank handle in a counterclockwise direction (fig. 41).

b. To Emplace the Mount.

(1) Unlimber the prime mover from the drawbar.

NOTE: The operation of disconnecting both bogies is identical.

(2) Operate the winch until the chain takes all the weight from the locking jaws (fig. 42).

(3) While one man steadies the winch, disengage one locking jaw at a time by raising handle (figs. 42 and 43). Repeat for the other locking jaw on the bogie.

OPERATION

RA PD 71214

RAMMER GUARD IN
OPERATING POSITION

Figure 41 — Cocking Automatic Rammer

47

GERMAN 88-MM ANTIAIRCRAFT GUN MATERIEL

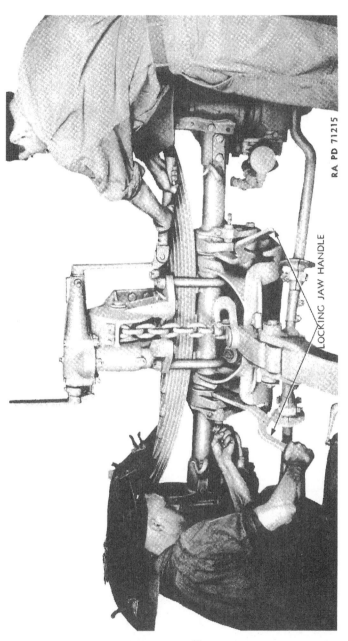

RA PD 71215

LOCKING JAW HANDLE

Figure 42 — Unlocking Front Bogie from Bottom Carriage

OPERATION

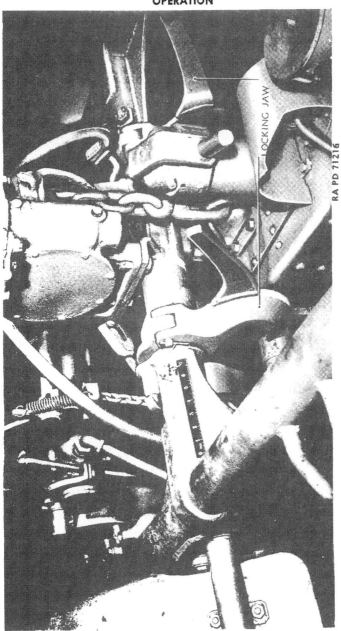

LOCKING JAW

RA PD 71216

Figure 43 — Rear Bogie Locking Jaws

49

GERMAN 88-MM ANTIAIRCRAFT GUN MATERIEL

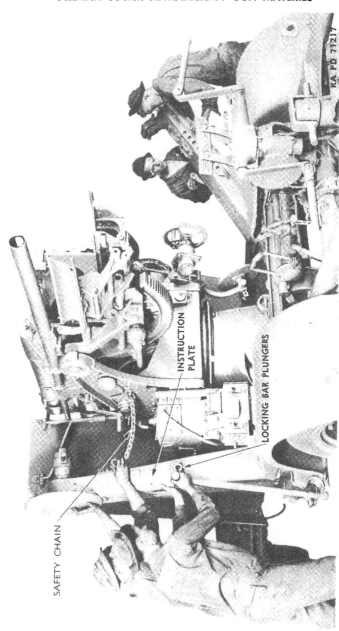

Figure 44 — Releasing Locking Bar Plunger and Removing Safety Chain

OPERATION

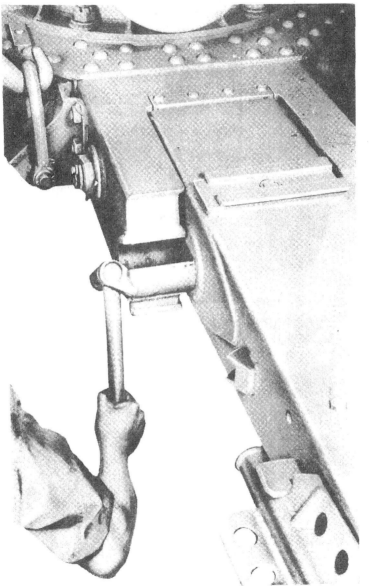

Figure 45 — Locking Outriggers in Place

GERMAN 88-MM ANTIAIRCRAFT GUN MATERIEL

BOGIE CHAIN

RA PD 71219

Figure 46 — Mount Lowered to Ground

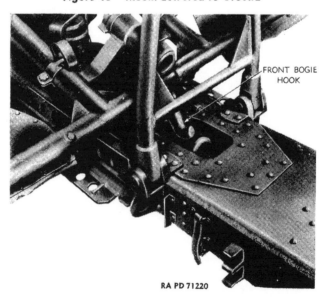

FRONT BOGIE
HOOK

RA PD 71220

Figure 47 — Front Bogie Hook Disengaged

52

OPERATION

REAR BOGIE HOOKS

RA PD 71221

Figure 48 — Rear Bogie Hooks Engaged

RA PD 71222

Figure 49 — Supporting Mount and Outriggers on Leveling Jacks

GERMAN 88-MM ANTIAIRCRAFT GUN MATERIEL

RA PD 71223

Figure 50 — Staking Mount and Outriggers

(4) Simultaneously with the above, lower the side outriggers by performing the following steps. NOTE: The instruction plate (fig. 44) on the left outrigger reads "VOR AUSLOSEN DER STUTZEN SEITENHOLM DURCH 2 MANN FESTHALTEN," which translated means, "Before releasing the side outrigger supports, steady (the outriggers) by 2 men."

(a) Releasing the locking bar plunger (fig. 44).

(b) Removing the safety chains (fig. 44).

(c) Swinging the outriggers down.

(d) Locking the outriggers in place by rotating the locking pins (fig. 45).

(5) When the mount is completely lowered, unhook the bogie chains (fig. 46).

(6) Disengage the hooks securing the bogies to the mount (figs. 47 and 48). NOTE: Unhook the front bogie first.

(7) Remove the bogies, connect them together with the transporter bar, and wheel them away as a complete trailer unit.

(8) Engage the elevating gear clutch by pulling the clutch lever to its downward position (fig. 37).

(9) Release the muzzle rest (fig. 38) by:

(a) Unscrewing the muzzle rest lock so that the chain may be swung over the barrel.

OPERATION

(b) Elevating the gun slightly so that the muzzle rest may be pushed forward and down onto the bottom carriage.

(10) Support the bottom carriage and side outriggers with the leveling jacks (fig. 49).

(11) Secure the mount in position by driving the stakes through the bottom carriage and outriggers as shown in figure 50.

(12) Level the top carriage by the cross-leveling handwheels, using the level indicator for reference (fig. 39).

(13) Unfold the rammer guard (fig. 40) to the operating position (fig. 41).

(14) Cock the rammer assembly by rotating the rammer crank handle in a counterclockwise direction (fig. 41).

22. TO TRAVERSE.

a. The traversing handwheel is located on the right side of the carriage (fig. 10). Rotate the handwheel clockwise for right traverse and counterclockwise for left traverse, either in high or low speed.

(1) To TRAVERSE IN LOW SPEED. Trip the gear selector lever (fig. 31) toward the mount so that the lever will engage one of the four notches on the inner collar on the traversing handwheel shaft.

(2) To TRAVERSE IN HIGH SPEED. Trip the gear selector lever away from the mount so that the lever will engage one of the four notches on the outer collar on the traversing handwheel shaft.

23. TO ELEVATE.

a. The elevating handwheel is located on the right side of the carriage (fig. 10). Rotate the handwheel clockwise for elevation and counterclockwise for depression, either in high or low speed.

(1) To ELEVATE IN HIGH SPEED. Trip the gear selector lever away from the mount so that the lever will engage one of the four notches on the outer collar on the elevating handwheel shaft.

(2) To ELEVATE IN LOW SPEED. Trip the gear selector lever toward the mount so that the lever will engage one of the four notches on the inner collar on the elevating handwheel shaft.

24. TO OPERATE THE BREECH MECHANISM.

a. To Open.

(1) Normally, in action, the breech is opened, percussion mechanism cocked, and cartridge case extracted during counterrecoil of the gun.

(2) To open the breech manually before inserting the initial round of ammunition, grasp the breech actuating lever and squeeze the trigger to release the retaining catch (fig. 51). Rotate the breech actuating lever clockwise as far as it will go.

(3) Opening the breech manually may be performed either during the engaged or disengaged position of the "SEMIAUTOMATIC-

GERMAN 88-MM ANTIAIRCRAFT GUN MATERIEL

A — CATCH PLUNGER   E — TRIGGER
B — CATCH   F — BREECH ACTUATING LEVER
C — BREECH ACTUATING MECHANISM   G — EXTRACTOR ACTUATING LEVER
D — {COCKING LEVER IN "FIRE"
      /"FEUER" POSITION

RA PD 71224

Figure 51 – Firing the Gun Manually

OPERATION

RA PD 71225

Figure 52 — Disengaging "SEMIAUTOMATIC-HAND" Catch

HAND" ("SEMIAUTOMATIK-HAND") catch. With the catch engaged, a strong pull to rotate the breech actuating lever is necessary. To engage the catch, pull down on the catch plunger and raise the catch in front to engage the breech actuating mechanism (fig. 51). The catch is disengaged when it is pressed down in front (fig. 52).

b. To Close.

(1) Normally, in action, the breech is closed by the cartridge base tripping the extractors, thereby releasing the breechblock, which closes due to the force of the spring in the breech actuating mechanism.

(2) After the firing period, it is necessary to close the breech. This is accomplished by rotating the extractor actuating lever in a clockwise direction (fig. 51) or operating the loading tray interlock without a round in the tray.

25. POINTS TO BE OBSERVED BEFORE FIRING.

a. **Lubrication.** All points should be thoroughly lubricated as prescribed (par. 33). The recoil, recuperator, and rammer cylinders should be filled to proper oil levels (pars. 37, 38, and 39). The recuperator and rammer cylinders should be charged to proper gas pressure (pars. 37 and 39).

GERMAN 88-MM ANTIAIRCRAFT GUN MATERIEL

26. POINTS TO BE OBSERVED DURING FIRING.

a. If the gun fails to fire, the following safety precautions must be observed:

(1) Stand clear of the path of recoil.

(2) Keep the gun at firing elevation. Do not depress the piece.

(3) Keep the gun directed in traverse either on the target or on a safe place in the field of fire.

(4) The breech will not be opened until at least 10 minutes after the last unsuccessful attempt to fire the piece.

27. TO LOAD.

a. Place the shell on the loading tray and swing the tray in line with the axis of the bore of the gun. At this point the loading tray interlock is released and the expanding gas forces the rammer cylinder back along the piston; thus the rammer arm is rapidly withdrawn, seating the round (fig. 53). Swing the empty loading tray back to its original outboard or loading position.

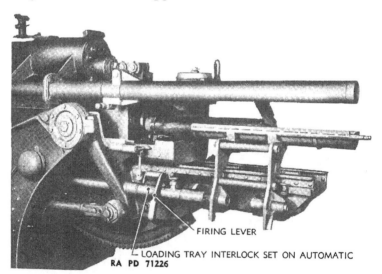

FIRING LEVER

LOADING TRAY INTERLOCK SET ON AUTOMATIC
RA PD 71226

Figure 53 — Shell Partially Rammed

b. When firing at angles above 45 degrees, set the buffer valve to "QUICK" ("SCHNELL") (fig. 35) by turning the valve clockwise. At angles below 45 degrees, the valve is set to "NORMAL" ("NORMAL") by rotating the valve counterclockwise.

OPERATION

28. TO FIRE.

a. With the loading tray interlock set at "AUTOMATIC" ("AU-TOMATIK") (fig. 53), the gun will fire as soon as the loading tray clears the path of recoil and is returned by hand to its outboard or firing position.

b. With the loading tray interlock set at "HAND" ("HAND"), the gun must be fired by performing either one of the following steps:

(1) Raising the firing lever on the left side of the cradle (fig. 53).

(2) Pulling the auxiliary trigger on the right side of the cradle (fig. 51).

29. TO RECOCK.

a. In case of a misfire, it will be necessary to recock the percussion mechanism by rotating the cocking lever in a counterclockwise direction as far as the word "WEIDERSPANNEN," which means "RE-COCK." Then return the cocking lever to its original position at "FIRE" ("FEUER") (fig. 51). Fire the gun as described in paragraph 28 again; and if the gun again misfires, wait 10 minutes and then unload as described in paragraph 30.

30. TO UNLOAD.

a. Open the breech. If the extractor does not eject the shell, grasp the shoulder on the cartridge base and withdraw it from the breech recess. Then reload the gun.

31. TO PLACE THE WEAPON IN TRAVELING POSITION.

a. To place the weapon in traveling position after having been fired from the wheels:

(1) Release the pressure on the rammer by operating the loading tray interlock without a round in the tray. The rammer guard may now be folded back.

(2) Swing the muzzle rest up to vertical position and secure the gun to it.

(3) Disengage the elevating gear clutch.

(4) Release the hand brakes.

(5) Connect the prime mover to the drawbar.

b. To place the weapon in traveling position after having been fired from emplacement:

(1) Release the pressure on the rammer by operating the loading tray interlock without a round in the tray. The rammer guard may now be folded back.

(2) Withdraw the stakes and secure them in their places on the outriggers.

(3) Swing the muzzle rest up to vertical position and secure the gun to it.

(4) Disengage the elevating gear clutch.

(5) Remove the transporter bar from between the bogies.

GERMAN 88-MM ANTIAIRCRAFT GUN MATERIEL

(6) Secure the bogies to the mount by engaging the hooks provided for the purpose.

(7) Connect the bogie chains to the bottom carriage.

(8) Place the side outriggers in traveling position by:

(a) Unlocking the outriggers by rotating the locking pins.

(b) Swinging the outriggers up against the mount.

(c) Engaging the locking bar plungers.

(d) Securing the safety chains.

(9) Operate the winch to raise the bottom carriage high enough to engage the locking jaws.

(10) Lower the bottom carriage until its weight settles in the locking jaws.

NOTE: Use steps (9) and (10) with the front bogie first.

(11) Connect the prime mover to the drawbar.

CHAPTER 2

GERMAN 88-MM ANTIAIRCRAFT GUN AND MOUNT (Cont'd)

Section V

LUBRICATION

32. INTRODUCTION.

a. Lubrication is an essential part of preventive maintenance, determining to a great extent the serviceability of parts and assemblies.

33. LUBRICATION GUIDE.

a. **General.** Lubrication instructions for this materiel are consolidated in the lubrication guides (figs. 54 and 55). These specify the points to be lubricated, the periods of lubrication, and the lubricant to be used. In addition to the items on the guides, other small moving parts, such as hinges and latches, must be lubricated at frequent intervals.

b. **Supplies.** In the field it may not be possible to supply a complete assortment of lubricants called for by the lubrication guides to meet the recommendations. It will be necessary to make the best use of these available, subject to inspection by the officer concerned, in consultation with responsible ordnance personnel.

c. Oilholes and lubrication fittings are painted red for easy identification.

d. American lubrication guns and couplets will fit most German lubrication fittings.

e. American and German lubrication fittings are interchangeable.

f. All gear cases should be disassembled, cleaned, and lubricated with GREASE, O.D. (seasonal grade), by ordnance personnel at the earliest opportunity available, and every 6 months thereafter.

g. **Wheel Bearings.** Remove bearing cone assemblies from hub and wash spindle and inside of hub with SOLVENT, dry-cleaning. Wet the spindle and inside of hub and hub cap with GREASE, general purpose, No. 2, to a maximum thickness of $\frac{1}{16}$ inch only to retard rust. Wash bearing cones with SOLVENT, dry-cleaning. Inspect and replace if necessary. Lubricate bearings with GREASE, general purpose, No. 2, with a packer or by hand, kneading lubricant into all spaces in the bearing. Use extreme care to protect bearings from dirt, and immediately reassemble and replace wheel. Do not fill hub or hub cap. The lubricant in the bearings is sufficient to provide lubrication until the next service period. Any excess might result in leakage into the brake drum.

GERMAN 88-MM ANTIAIRCRAFT GUN MATERIEL

RA PD 71227

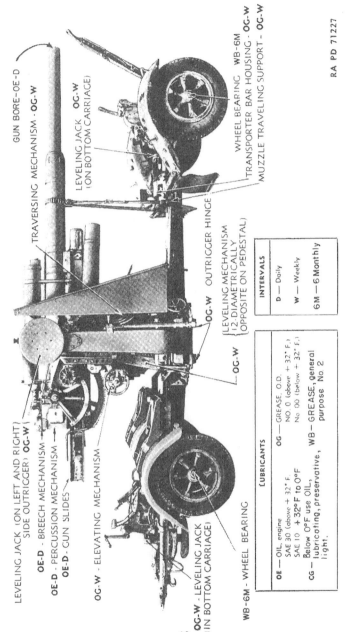

LUBRICANTS

OE — OIL, engine	**OG** — GREASE, O.D.
SAE 30 (above + 32° F.)	NO. 0 (above + 32° F.)
SAE 10 + 32° F to 0°F	No. 00 (below + 32° F.)
Below 0°F use OIL,	**WB** — GREASE, general
lubricating, preservative,	purpose No 2
light.	

INTERVALS

D — Daily	
W — Weekly	
6M — 6 Monthly	

Figure 54 — Lubrication Guide

62

LUBRICATION

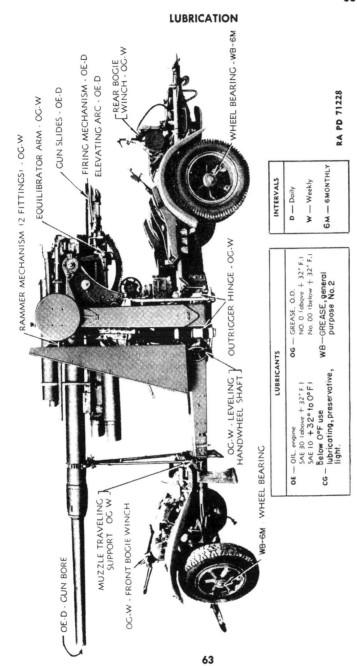

Figure 55 — Lubrication Guide

RA PD 71228

CHAPTER 2

GERMAN 88-MM ANTIAIRCRAFT GUN AND MOUNT (Cont'd)

Section VI

CARE AND PRESERVATION

34. GENERAL.

a. Gun.

(1) The cannoneer will examine the bore before each loading to ascertain and remove, if necessary, portions of powder bag or unburned powder remaining in the bore.

(2) In cleaning after firing, wash the bore with a solution of ½ pound of SODA ASH in 1 gallon of water. Wipe dry with the bore sponge covered with clean white rags. Oil the bore with OIL, engine, SAE 10, if temperature is between plus 32 F and 0 F. Use OIL, engine, SAE 30, above plus 32 F.

(3) Lubricating instructions are given in paragraph 33.

(4) When the materiel is not in use, covers must be used.

b. Breech Mechanism. The breech mechanism should be kept clean and the parts well lubricated. Disassemble daily or after firing, clean with SOLVENT, dry-cleaning, and oil with OIL, engine (seasonal grade).

c. Firing Mechanism. These parts require the same attention as the breech mechanism. Therefore, frequent disassembly for the purpose of cleaning and lubrication according to the lubrication guides (figs. 54 and 55), is required.

35. MOUNT.

a. Attention should be given to cleaning, lubricating, and to loose or broken parts. Lubrication, with the method and frequency of application, is carried in detail in paragraph 33.

h. The mount should be given a daily general inspection by the chief of section of the gun crew.

CARE AND PRESERVATION

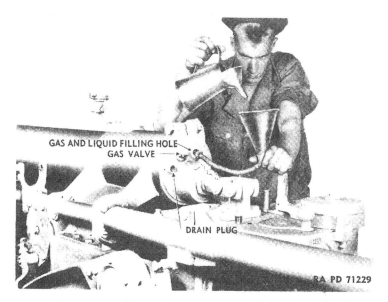

Figure 56 — Filling the Recuperator Cylinder with Liquid

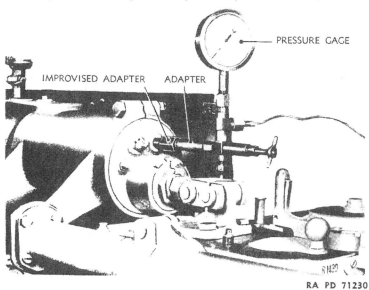

Figure 57 — Charging Recuperator Cylinder with Nitrogen

GERMAN 88-MM ANTIAIRCRAFT GUN MATERIEL

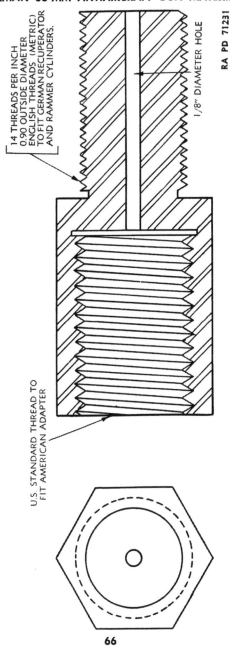

14 THREADS PER INCH
0.90 OUTSIDE DIAMETER
ENGLISH THREADS (METRIC)
TO FIT GERMAN RECUPERATOR
AND RAMMER CYLINDERS.

1/8" DIAMETER HOLE

RA PD 71231

U.S. STANDARD THREAD TO
FIT AMERICAN ADAPTER

66

Figure 58 — Improvised Adapter

CARE AND PRESERVATION

36. RECOIL LIQUID.

a. The recuperator, recoil, and rammer cylinders are filled with the same liquid. The liquid used is 60 percent glycerine and 40 percent distilled water by volume with 1 ounce of caustic soda per 3 gallons of liquid. This mixture, while not a duplicate of the German liquid, is entirely satisfactory as a substitute. If the recommended liquid is not available, COMPOUND, antifreeze (ethylene glycol type), may be used in the same proportion as glycerine with the same gas pressures as when using liquid.

37. FILLING AND CHARGING THE RECUPERATOR CYLINDER.

a. Remove the liquid and gas filling plug and the drain plug at the rear end of the recuperator cylinder. Open the gas valve at least three full turns (fig. 56).

b. Place the gun at a zero-degree elevation and a zero-degree cant. Using a funnel, pour in recoil liquid at the gas and liquid filling hole (fig. 56) until it overflows at the drain hole. Approximately 4½ gallons will be required. Replace the drain plug.

c. Screw adapter and pressure gage into the gas and liquid filling hole (fig. 57) and connect to nitrogen supply. NOTE: To make the American adapter fit the liquid and gas filling hole in the recuperator cylinder, as shown in figure 57, an adapter must be improvised. The male threads at one end of this improvised adapter are 14 threads per inch and the outside diameter is 0.900 English thread. The female threads will receive the American adapter. An ⅛-inch diameter hole runs through this adapter to allow passage of the gas (fig. 58).

d. Close the gas valve on the recuperator cylinder and check the gas line for leakage.

e. If the gas line is tight, open the gas valve on the recuperator cylinder about two turns and charge with gas until approximately 600 pounds per square inch are recorded on the pressure gage. Close the gas valve on the recuperator cylinder and disconnect the gas line. Replace the gas and liquid filling plug.

38. FILLING THE RECOIL CYLINDER.

a. Elevate the gun to approximately a 2-degree elevation and a 0-degree cant.

b. Remove the two liquid filling plugs at the top front end of the recoil cylinder. Also remove the two overflow plugs, one at the front end and one at the left side of the cylinder.

c. Using a funnel, pour in recoil liquid at one of the liquid filling holes (fig. 59) until it overflows at the front overflow plug (fig. 59). Approximately 2½ gallons are required. Replace the front overflow plug.

d. Continue to pour until the liquid overflows at one of the top filling holes. As entrapped gas will cause the liquid to overflow, it is

GERMAN 88-MM ANTIAIRCRAFT GUN MATERIEL

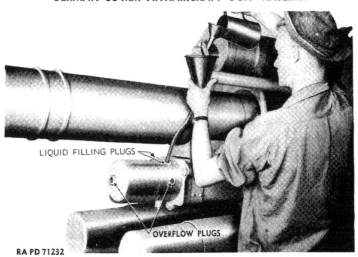

LIQUID FILLING PLUGS

OVERFLOW PLUGS

RA PD 71232

Figure 59 — Filling Recoil Cylinder with Liquid

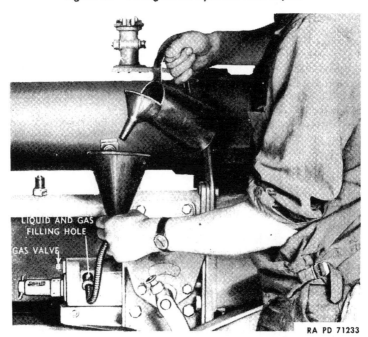

LIQUID AND GAS FILLING HOLE

GAS VALVE

RA PD 71233

Figure 60 — Filling Rammer Cylinder with Liquid

CARE AND PRESERVATION

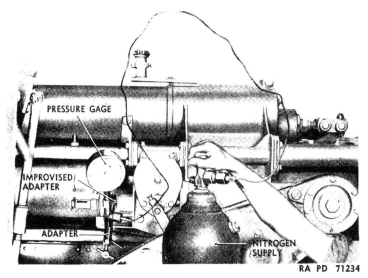

RA PD 71234

Figure 61 — Charging Rammer Cylinder with Nitrogen

desirable to wait until the turbulence subsides; then continue to fill. Replace the liquid filling plugs.

e.   Depress the gun to the horizontal and pour recoil liquid in the side overflow hole until it overflows. Replace the plug.

f.   Elevate and fully depress the piece at least three times to free the entrapped gas, and then elevate to a 2-degree elevation. Remove the liquid filling plugs again and add liquid if necessary.

39.   FILLING AND CHARGING THE RAMMER CYLINDER.

a.   Place the gun at a zero-degree elevation and a zero-degree cant.

b.   Remove the liquid and gas filling plug.

c.   Open the gas valve about two turns (fig. 60).

d.   Using a funnel, add recoil liquid until the liquid overflows at the gas and liquid filling hole.

e.   Screw adapter and pressure gage into the gas and liquid filling hole (fig. 61) and connect to nitrogen supply. NOTE: Use the same improvised adapter described in paragraph 37 c (fig. 58).

f.   Close the gas valve on the rammer cylinder and check the gas line for leakage.

g.   If the gas line is tight, open the gas valve on the rammer cylinder about two turns and fill with nitrogen until approximately 225 pounds per square inch are recorded on the pressure gage. Close the gas valve fully and disconnect the gas line. Replace the gas and liquid filling plug.

GERMAN 88-MM ANTIAIRCRAFT GUN MATERIEL

CHAPTER 2

GERMAN 88-MM ANTIAIRCRAFT GUN AND MOUNT (Cont'd)

Section VII

INSPECTION AND ADJUSTMENT

40. GENERAL.

a. Inspection has as its purpose the detection of conditions which cause improper performance. Such conditions may be caused by:

(1) Mechanical deficiencies resulting from ordinary wear and breakage.

(2) Faulty or careless operation.

(3) Improper care (servicing and lubrication). Inspection should always be accompanied by corrective measures to remedy any deficiencies found. When properly carried out, inspection and necessary corrective maintenance will insure the maximum reliability and performance of the materiel. The following inspection should be made at regular intervals not to exceed 30 days during both active and inactive seasons.

b. Before making a detailed inspection, the gun and mount should be inspected in general for evidences of faulty operation, care, and maintenance. Any unusual conditions which might result in improper operation or damage to the materiel, must be immediately remedied. Untidy appearance and evidences of rust or deterioration must be corrected. Missing or broken apparatus must be replaced.

41. INSPECTION OF GUN.

Inspection	Adjustment
Note condition of bore. Look for rust. Note lubrication of bore.	Remove any rust. Clean and slush. Wipe dry and lubricate.

INSPECTION AND ADJUSTMENT

42. INSPECTION OF BREECH MECHANISM.

Inspection

Note smoothness of operation of the breech mechanism in opening and closing.

Adjustment

If the mechanism does not operate smoothly, disassemble, clean, examine the parts for wear or breakage, and replace unserviceable parts. Lubricate the mechanism and reassemble. If it is still difficult to operate, notify ordnance maintenance personnel.

Examine the breechblock and breech recess for burs, indentations, rust, pitting, and other evidence of erosion.

If not possible to smooth or clean with CLOTH, crocus, notify ordnance maintenance personnel. Do not use any other abrasive.

43. INSPECTION OF FIRING MECHANISM.

Note the action of the firing mechanism by pulling on the firing lever. Then open the breech and note whether the percussion mechanism has been cocked during the operation of opening the breech. Close the breech again and operate the firing mechanism. With the breech closed, cock the percussion mechanism using the cocking lever assembly.

Repair or replace parts of the percussion mechanism release assembly, percussion mechanism, and cradle firing mechanism which may be worn or broken.

44. INSPECTION OF TRAVERSING MECHANISM.

Operate the traversing mechanism to determine smoothness of operation and whether there is any backlash or play in the mechanism.

Lubricate. If this fails, notify ordnance maintenance personnel.

Inspect for defective or broken parts.

Notify ordnance maintenance personnel.

Examine for proper lubrication.

Lubricate if necessary.

45. INSPECTION OF ELEVATING MECHANISM.

Operate the elevating mechanism to determine smoothness of operation and whether

Lubricate. If this fails, notify ordnance maintenance personnel.

Inspection	Adjustment
there is any backlash or play in the mechanism.	
Inspect for defective or broken parts.	Notify ordnance maintenance personnel.
Examine for proper lubrication.	Lubricate if necessary.

46. INSPECTION OF RECOIL AND RECUPERATOR MECHANISMS.

Inspection	Adjustment
Check recuperator cylinder for proper amount of gas pressure and liquid.	To check the gas pressure:

To check the gas pressure:

See that the gas valve on the recuperator cylinder is closed.

Remove the liquid and gas filling plug and install the pressure gage securely.

Open the gas valve and read the pressure registered on the gage. The normal pressure is 600 pounds per square inch. If the recorded pressure is less than normal, proceed as in paragraph 37.

The gas pressure may also be checked without the use of the gage, by elevating the gun to maximum elevation and then jacking it out of battery and blocking in the recoil position. If the gun slides into battery rapidly when the block is knocked out, the gas pressure is sufficiently high for proper operation.

To check the liquid level:

Depress the gun to minus 1 degree.

Gently open drain plug not more than one turn and note whether liquid flows. If it does, there is sufficient liquid.

If no liquid flows, the cylinder must be purged of gas and refilled with liquid as described in paragraph 37.

INSPECTION AND ADJUSTMENT

Inspection	Adjustment
Check the recoil cylinder for proper amount of liquid.	To check liquid: Elevate the gun to a 2-degree elevation. Remove the liquid *filling* plugs. The liquid should be up to the level of the filling holes. If it is necessary to add liquid, proceed as in paragraph 38.
Exercise the weapon using a block and tackle to determine the amount of recoil, which should be 41½ inches at a 0-degree elevation and 27¾ inches at maximum elevation.	If necessary, check the gas and liquid content of the recuperator cylinder, should the weapon recoil in excess of the prescribed distances (par. 46 a). If cylinder is full, notify ordnance maintenance personnel.
The gun should not jump or slam into battery. The gun should ease into battery smoothly against the action of the recoil mechanism.	If necessary, refill the recoil cylinder. If this does not remedy the condition, notify ordnance maintenance personnel.
Note whether or not the recoil and recuperator piston rods are properly secured to the breech ring.	Tighten if necessary.
Inspect for any leakage of liquid around the recoil or recuperator cylinders.	Notify ordnance maintenance personnel.

47. INSPECTION OF RAMMER ASSEMBLY.

Check the rammer cylinder for proper amount of gas pressure and liquid.	To check the gas pressure: See that the gas valve in the rammer cylinder is closed. Remove the liquid and gas filling plug at the air filling vent and install the pressure gage securely. Open the valve and read the pressure registered on the gage. The normal pressure is 225 pounds per square inch. If the recorded pressure is less than normal, proceed as in paragraph 39.

Inspection	Adjustment
	To check the liquid level in the rammer cylinder.
	Depress the gun to minus 1 degree.
	Gently open the gas and liquid filling plug not more than one turn and note whether liquid flows. If it does, there is sufficient liquid.
	If no liquid flows, the cylinder must be purged of gas and refilled with liquid as described in paragraph 39.
Check the smoothness of the rammer tray for burs or rust.	If not possible to smooth or clean with CLOTH, crocus, notify ordnance maintenance personnel. Do not use any other abrasive.
Inspect for any leakage around the rammer cylinder.	Notify ordnance maintenance personnel.
Operate the rammer assembly by cocking the mechanism; then release the pressure on the rammer to determine smoothness of operation.	Lubricate if necessary. If this fails, notify ordnance maintenance personnel.

48. INSPECTION OF MOUNT.

Inspection	Adjustment
Inspect the pintle bearing for lubrication.	Lubricate if necessary.
Inspect the trunnion bearings for cleanliness and lubrication.	Clean and lubricate.

49. INSPECTION OF BOGIES.

Inspection	Adjustment
Examine the winches for broken parts, smoothness of operation, and lubrication.	Repair any damage or replace broken parts. Lubricate if necessary.
Check to see that the leaf spring clips are tight and that the spring center bolts are not worn.	Tighten the spring clips or replace spring center bolts, if necessary.

50. INSPECTION OF BRAKES.

a. **Power Brakes.** Inspect the power brake mechanism at frequent intervals to discover air leaks. All air line connections must be tight.

INSPECTION AND ADJUSTMENT

In case leakage test shows a 2-inch diameter soap bubble in 5 seconds, notify ordnance maintenance personnel.

b.. **Hand Brakes.** A hand brake lever is mounted on the rear bogie. The lever is retained in position by a latch engaging a toothed segment. If the hand brake does not hold, notify ordnance maintenance personnel for any necessary adjustment.

51. INSPECTION OF EQUILIBRATORS.

a. If the elevation handwheel is difficult to operate, it is possible that the equilibrators are not compensating for the unbalanced weight of the gun. Notify ordnance maintenance personnel for any adjustment necessary.

GERMAN 88-MM ANTIAIRCRAFT GUN MATERIEL

CHAPTER 2

GERMAN 88-MM ANTIAIRCRAFT GUN AND MOUNT (Cont'd)

Section VIII

MALFUNCTIONS AND CORRECTIONS

52. MALFUNCTION OF GUN.

a. Fails to Fire; No Percussion on Primer.

Cause	Correction
Broken or weak firing spring. Broken or deformed firing pin.	Remove firing spring retainer, firing spring, and firing pin holder assembly. Replace broken or deformed parts. Clean and lubricate; then replace in breechblock.
Sear not retaining the firing pin in cocked position.	Remove the sear and sear spring. Clean and lubricate; then replace.

b. Fails to Fire Until After Several Percussions on Primer.

Cause	Correction
Percussion mechanism and percussion mechanism release assembly parts not working freely.	Disassemble and examine carefully for burs, or rough surfaces. Smooth with CLOTH, crocus, or an oil stone. Clean, lubricate, and reassemble.
Weak firing spring.	Replace.

c. Fails to Fire When Proper Pressure on Primer is Obtained.

Cause	Correction
Defective primer.	After three percussions, wait 2 minutes before opening breech; then insert another round of ammunition.

d. Fails to Extract Empty Cartridge Case.

Cause	Correction
Broken extractor.	Carefully remove the case by operating from the muzzle end. Examine the edge of the chamber for deformation or burs which might cause difficult extraction. Disassemble mechanism. Replace extractor, if necessary.

MALFUNCTIONS AND CORRECTIONS

e. **Misfire.**

Cause	Correction
Defective ammunition.	In case of a misfire, at least two or three additional attempts to fire the piece should be made. The breechblock will not be opened until at least 10 minutes after the last unsuccessful attempt to fire the piece. The gun will be kept directed in elevation and traverse either on the target or on a safe place in the field of fire.

f. **Breechblock Unable to Be Brought to Full Closed Position.**

Improper chambering of cartridge case.	Attempt to close the breech. If the breech will not close, open the breech and insert another round. If the malfunction recurs, notify the ordnance maintenance personnel.
Breechblock seized.	Notify ordnance maintenance personnel.

53. MALFUNCTION OF MOUNT.

a. **Gun Returns to Battery with Too Great a Shock.**

Excessive amount of recoil liquid in recuperator cylinder.	Drain and recharge the recuperator cylinder with liquid and air pressure, as described in paragraph 37.
Insufficient amount of liquid in recoil mechanism.	Fill recoil mechanism, as described in paragraph 38.
Recoil mechanism out of order.	Notify ordnance maintenance personnel.

b. **Gun Fails to Return to Battery.**

Excessive friction as stuffing boxes.	Notify ordnance maintenance personnel.
Damaged recoil slides, piston rod, or piston.	Notify ordnance maintenance personnel.
Insufficient amount of liquid and gas pressure in recuperator.	Charge the recuperator with liquid and gas pressure, as described in paragraph 37.
Recoil mechanism out of order.	Notify ordnance maintenance personnel.

GERMAN 88-MM ANTIAIRCRAFT GUN MATERIEL

Cause	Correction
Lack of lubrication, or scoring or "freezing" of sliding surfaces.	Notify ordnance maintenance personnel.

c.   Spasmodic Counterrecoil.

Lack of lubrication or scoring of sliding surfaces.	Notify ordnance maintenance personnel.

d.   Gun Recoils More Than the Maximum Distance Allowed.

Insufficient amount of liquid and gas pressure in recuperator cylinder.	Charge the recuperator cylinder with liquid and gas pressure, as described in paragraph 37.

e.   Gun Slides Out of Battery When Slightly Elevated.

Insufficient amount of liquid and gas pressure in recuperator cylinder.	Charge the recuperator cylinder with liquid and gas pressure, as described in paragraph 37. If trouble repeats, notify ordnance maintenance personnel.

CHAPTER 2

GERMAN 88-MM ANTIAIRCRAFT GUN AND MOUNT (Cont'd)

Section IX

DISASSEMBLY AND ASSEMBLY

54. GENERAL.

a. Wear, breakage, cleaning, and inspection, make necessary the occasional disassembly of various parts of the gun and mount. This work comes under two headings, that which may be performed by the battery personnel with the equipment furnished and that which must be performed by trained ordnance personnel.

b. The battery personnel may, in general, do such dismounting as is required for battery use. Such work should be done in the manner prescribed herein. Any difficulty which cannot be overcome by the pre- scribed method must be brought to the attention of ordnance personnel.

c. The battery personnel will not attempt to disassemble any part of the recoil mechanism not authorized in this manual, nor do any filing on the sights or gun parts; and only by order of the battery commander on any mount part.

d. The use of wrenches that do not fit snugly on the parts should be avoided. They will not only fail to tighten the part properly but will damage the corners of the nuts and bolt heads. There is also danger of spreading the wrenches and rendering them useless.

e. Before attempting the assembly of the larger mechanisms, the assembly of the subassemblies should be completed. In all assembly operations, the bearings, sliding surfaces, threads, etc., should be cleaned and lubricated.

55. TO DISASSEMBLE THE BREECH MECHANISM.

a. Set the gun at a 0-degree elevation and close the breech. Re- move the locking pin (fig. 13) and remove the breech operating crank (fig. 62).

b. Disengage the "SEMIAUTOMATIC-HAND" ("SEMIAUTO- MATIK-HAND") catch from the breech actuating mechanism (fig. 52). Then remove the breech actuating mechanism as a unit (fig. 63).

c. Rotate the extractor actuating lever and remove it from the breech ring (fig. 64).

GERMAN 88-MM ANTIAIRCRAFT GUN MATERIEL

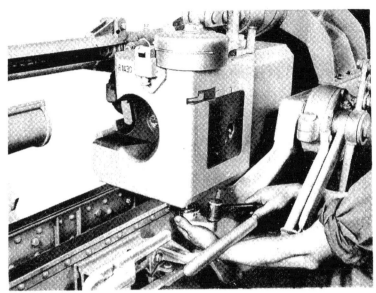

RA PD 71235

Figure 62 — Removing Breech Operating Crank

RA PD 71236

Figure 63 — Removing Breech Actuating Mechanism

DISASSEMBLY AND ASSEMBLY

RA PD 71237

Figure 64 — Removing Extractor Actuating Lever

RA PD 71238

Figure 65 — Removing Breechblock Actuating Lever

GERMAN 88-MM ANTIAIRCRAFT GUN MATERIEL

Figure 66 — Removing Extractor   RA PD 71239

RA PD 71240

Figure 67 — Removing Cocking Lever

82

DISASSEMBLY AND ASSEMBLY

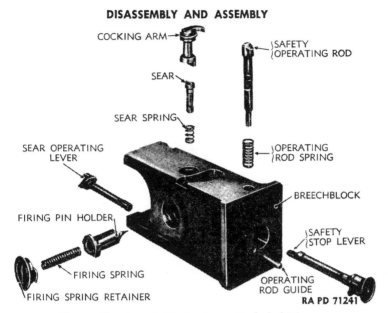

COCKING ARM

SAFETY
OPERATING ROD

SEAR

SEAR SPRING

OPERATING
ROD SPRING

SEAR OPERATING
LEVER

BREECHBLOCK

FIRING PIN HOLDER

SAFETY
STOP LEVER

FIRING SPRING

OPERATING
ROD GUIDE

FIRING SPRING RETAINER

RA PD 71241

Figure 68 — Breech Mechanism — Exploded View

d. Open the breech sufficiently and remove the breechblock actuating lever (fig. 65).

e. With the breechblock open 2 inches, press the heels of the extractors into the recesses of the breechblock.

f. Open breech sufficiently and remove the extractor (fig. 66).

g. Rotate the cocking lever counterclockwise until the handle is toward the muzzle. The lever may now be lifted out (fig. 67).

h. Remove the breechblock.

i. Fire the percussion mechanism by depressing the operating rod and rotating the sear operating lever (fig. 68).

j. Press in and rotate the firing spring retainer through 90 degrees. Remove the retainer and firing spring (fig. 68).

k. Rotate the cocking arm and remove the firing pin holder to the rear (fig. 68).

l. Lift out the cocking arm.

m. Depress the operating rod and remove the safety stop lever (fig. 68). The operating rod with spring may now be removed (fig. 68).

n. Press the sear down and remove the sear operating lever (fig. 68). The sear and spring may now be removed (fig. 68).

GERMAN 88-MM ANTIAIRCRAFT GUN MATERIEL

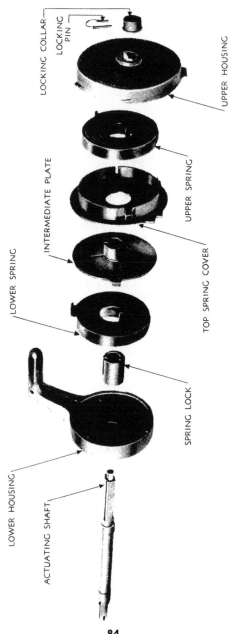

RA PD 71242

Figure 69 — Breech Actuating Mechanism — Exploded View

DISASSEMBLY AND ASSEMBLY

56. TO ASSEMBLE THE BREECH MECHANISM.

a. To assemble the breech mechanism, perform the steps necessary to disassemble the breech mechanism in the reverse order.

57. TO DISASSEMBLE THE BREECH ACTUATING MECHANISM.

a. With the breechblock removed, replace the breech actuating mechanism as a unit in the breech ring, with the breech mechanism actuating lever and breech operating crank assembled.

b. Engage the "SEMIAUTOMATIC-HAND" ("SEMIAUTO-MATIK-HAND") catch to the breech actuating mechanism.

c. Remove the locking pin and locking collar from the actuating shaft (fig. 69).

d. Rotate the breech mechanism actuating lever slightly towards the open position and remove the actuating mechanism upper housing (fig. 69). Allow the lever to rotate beyond the closed position until all spring pressure is lost. Remove the upper spring and top spring cover (fig. 69).

e. With the actuating mechanism in closed position, set the actuating lever and rotate the mechanism beyond the closed position and hold while the intermediate plate (fig. 69) is lifted up and removed; then gently ease the lever beyond the open position until all spring pressure is released. Remove the lower spring and the actuating mechanism lower housing.

f. Remove the spring lock (fig. 69).

58. TO ASSEMBLE THE BREECH ACTUATING MECHANISM.

a. To assemble the breech actuating mechanism, perform the steps necessary to disassemble the breech actuating mechanism in the reverse order.

GERMAN 88-MM ANTIAIRCRAFT GUN MATERIEL

CHAPTER 3

AMMUNITION

59. GENERAL.

a. Ammunition for the German 88-mm antiaircraft gun is similar to U. S. fixed 90-mm rounds (fig. 70). However, the German 88-mm rounds may be identified, as described in paragraph 64, by markings and appearance. The 88-mm multipurpose (8.8 cm. Flak 36), the 8.8 cm. Pak, and the 8.8 cm. Flak 18 guns are chambered alike and may use the same ammunition.

60. FIRING TABLES.

a. These are not available, except for the range table for firing the armor-piercing projectile in paragraph 64.

61. CLASSIFICATION.

a. The German 88-mm gun ammunition is classified according to type of projectile (Granate, Gr.) as high-explosive or armor-piercing. The high-explosive shell (Sprenggranate, Sprgr.) contains a relatively large charge of high explosive and any one of the following types of point fuzes:

(1) Combination superquick and delay fuze.

(2) Inertia-operated mechanical time fuze.

(3) Spring-wound mechanical time fuze.

NOTE: The armor-piercing projectile (Panzergranate, Pzgr.) is provided with an armor-piercing cap, to aid in penetration of armor plate, and a windshield, to improve the ballistic properties. It contains a relatively small explosive charge and a base-detonating fuze, having a tracer element in its base.

AMMUNITION

62. AUTHORIZED ROUNDS.

a. The following rounds may be found for use in the German 88-mm guns.

TABLE I. GERMAN 88-MM ROUNDS

Nomenclature of Complete Round*	Action of Fuze	Weight of Projectile as Fired (Pounds)	Muzzle Velocity (Feet per Second)
8.8 cm. Sprgr. Patr. L/4.5 (kz.) m. Zt. Z. S/30 (8.8 cm. fixed H.E. shell, with spring-wound mechanical time fuze)	Time (30 sec.)	20.06	2,690
8.8 cm. Sprgr. Patr. L/4.5 (kz.) m. Zt. Z. S/30 Fg¹ (8.8 cm. fixed H.E. shell, with inertia-operated mechanical time fuze)	Time (30 sec.)	20.06	2,690
8.8 cm. Sprgr. Patr. L/4.5 (kz.) m. A.Z. 23/28 (8.8 cm. fixed H.E. shell, with percussion fuze)	Superquick or delay (0.11 sec.)	20.34	2,690
8.8 cm. Pzgr. Patr. m. Bd. Z. (8.8 cm. fixed A.P.C. shell, with base-detonating fuze)	Nondelay	20.71	2,657

63. PREPARATION FOR FIRING.

a. Complete rounds, when they have been removed from their packing containers (par. 66), and their fuzes properly adjusted, are ready for firing.

b. Should it be necessary to fuze or unfuze projectiles, authorized personnel only will do this work. A spanner wrench labeled "A.Z. 23 and Zt. Z. S/30" should be used if available. The fuze setter for "A.Z. 23 and Zt. Z. S/30" may also be used to screw and unscrew fuzes.

c. Fuzes are adjusted for the desired action as described in paragraph 65.

64. DESCRIPTION OF ROUNDS.

a. General. The components of a complete round of German 88-mm ammunition are shown in figure 71. A comparison of the 88-mm armor-piercing and high-explosive complete rounds with a U.S. 90-mm high-explosive round is illustrated in figure 70. The double rotating band on the German 88-mm projectiles immediately distinguishes these from the U.S. round, as does the double 360-degree crimps of the cartridge case to the projectile. Markings and labels on

---

*For an explanation of German abbreviations, see paragraph 72.

GERMAN 88-MM ANTIAIRCRAFT GUN MATERIEL

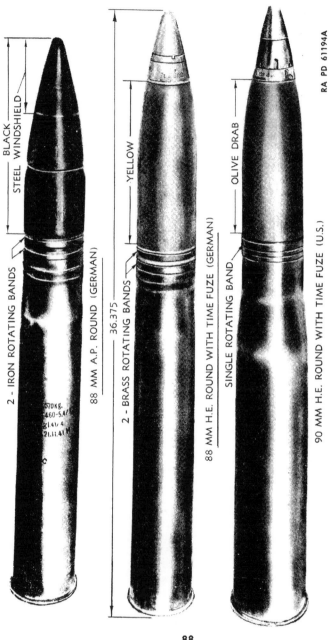

Figure 70 — Comparison of German 88-mm Armor-piercing Round, German 88-mm High-explosive Round, and U.S. 90-mm High-explosive Round with Time Fuze

AMMUNITION

shipping containers and crates (par. 66, figs. 77 and 78) serve as a means of identification. Further identification may be made by means of markings on the ammunition as described in subparagraph **b**, below, and is illustrated in figures 72 and 73.

b. Identification.

(1) GENERAL. The following identification markings may be found on projectiles. These markings may vary, dependent upon the particular lots found in the field. German abbreviations and terminology, and their English equivalents are given in paragraph 72.

(2) ON THE PROJECTILE. Armor-piercing projectiles are painted black above the rotating bands; high-explosive projectiles are painted yellow. In addition, the following markings may be present:

(a) Weight-zone Marking (Gewichtsklasse). The weight-zone marking is a roman numeral in black. The numeral "III" generally indicates "standard" weight; no weight corrections in the firing tables are necessary in firing shells which are in weight-zone III.

(b) Shell Number. In the case of 88-mm shell, the number 28.

(c) Date of assembly and manufacturer's initials or symbol.

(d) A number indicating type of high-explosive filler, for example: "1" indicates TNT; "2" indicates picric acid. Other number designations will be found in paragraph 72.

(e) Abbreviations denoting type of shell, for example:

1. Tp (Tropen), for the tropics.

2. Ub (Ubung), practice.

3. Nb (Nebel), smoke.

(3) ON THE SIDE OF THE CARTRIDGE CASE. Markings on the side wall of the cartridge case are shown in figure 72. Their English equivalents are given in Table II.

(4) ON THE BASE OF THE CARTRIDGE CASE. The principle marking for identification on the base of the cartridge case is the number "6347," which appears on all the cartridge cases of the German 88-mm complete rounds. For other markings, see figure 73. It will be noted that the primer design marking is "C/12nA St." The caliber and model of the gun may also appear on the base, for example, "8.8 cm. Flak. 18."

TABLE II. MARKINGS ON SIDE OF CARTRIDGE CASE

German Marking	English Equivalent
2.700 kg.	2,700 kilograms
Digl. R.P.—8 (495-5, 4/2.75)	Diglycol powder, number (grain size)
tgl. 41 P	Manufacturer, date, delivery number
Lu. 14, 5.41 K.	Manufacturer, date, work mark
P.T. plus 25 C (in red)	Powder temperature (pulvertempera-tur), plus 25 C

GERMAN 88-MM ANTIAIRCRAFT GUN MATERIEL

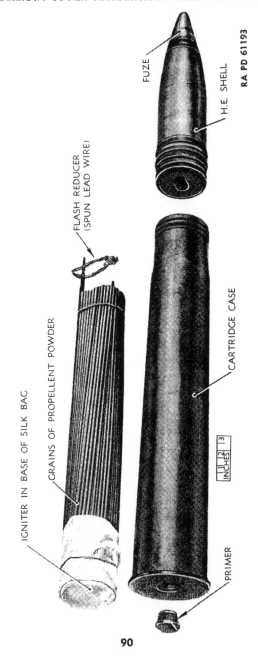

RA PD 61193

FUZE

H.E. SHELL

FLASH REDUCER
(SPUN LEAD WIRE)

GRAINS OF PROPELLENT POWDER

IGNITER IN BASE OF SILK BAG

CARTRIDGE CASE

INCHES 1 2 3

PRIMER

Figure 71 — Components of German 88-mm High-explosive Complete Round

AMMUNITION

c. **8.8 cm. Sprgr. Patr. L/4.5 (kz.) m. Zt. Z. S/30 (8.8 cm. High-explosive Fixed Shell, with Spring-wound Mechanical Time Fuze).**

(1) COMPLETE ROUND. This complete round, illustrated in figure 70, consists of cartridge case No. 6347, containing the primer and propelling charge, crimped to a high-explosive projectile which is fuzed with a 30-second spring-wound mechanical time fuze. It is identified as indicated in subparagraph b, above. The complete round weighs 31.69 pounds and is 36.69 inches in length. The maximum horizontal range is 16,200 yards, the vertical range being 32,500 feet. Muzzle velocity and weight of projectile are given in Table I. Packing of this round is described in paragraph 66.

(2) PROJECTILE. The two rotating bands on the projectile are fashioned from powdered iron impregnated with wax. The German high-explosive projectiles differ from those of U.S. manufacture, in that a base plug is screwed into the base. A lead gasket seals the base assembly against possible penetration of the hot gases of the exploded propellant. The German projectile has a square base and is 15.55 inches in length. The bursting charge consists of 2.19 pounds of TNT or 40/60 amatol.

(3) PREPARATION FOR FIRING. This round is ready for firing when removed from its packing, except that the mechanical time fuze must be set as described in paragraph 65.

d. **8.8 cm. Sprgr. Patr. L/4.5 (kz.) m. Zt. Z. S/30 Fg[1] (8.8 cm. Fixed High-explosive Shell with Inertia-operated Mechanical Time Fuze).** This complete round is the same as the fixed high-explosive round described in subparagraph c, above, except for the time fuze, which functions by different means. However, the timing of the fuze for this round is also 30 seconds. See paragraph 65 for description of fuzes.

e. **8.8 cm. Sprgr. Patr. L/4.5 (kz.) m. A.Z. 23/28 (8.8 cm. Fixed High-explosive Shell, with Percussion Fuze).** This complete round is the same as the fixed high-explosive round described in subparagraph c, above, except for the fuze, which is a combination superquick and delay (0.11 second) fuze similar in action and setting to the U.S. FUZE, P.D., M48, or FUZE, P.D., M51. See paragraph 65 for description of fuzes. This complete round weighs 32 pounds, the weight of the projectile being 20.34 pounds.

f. **8.8 cm. Pzgr. m. Bd. Z. (8.8 cm. Fixed Armor-piercing Capped Shell, with Base-detonating Fuze).**

(1) COMPLETE ROUND. This complete round, illustrated in figure 70, consists of cartridge case No. 6347, containing the primer and propelling charge, crimped to an armor-piercing projectile which contains a high-explosive filler, base-detonating fuze, and a tracer. It is identified as indicated in subparagraph b, above. The complete round weighs

GERMAN 88-MM ANTIAIRCRAFT GUN MATERIEL

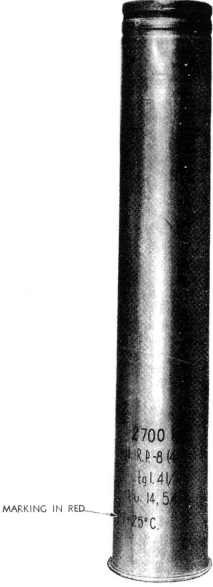

MARKING IN RED

RA PD 61197

Figure 72 — German 88-mm Cartridge Case, Showing Stenciled Markings

AMMUNITION

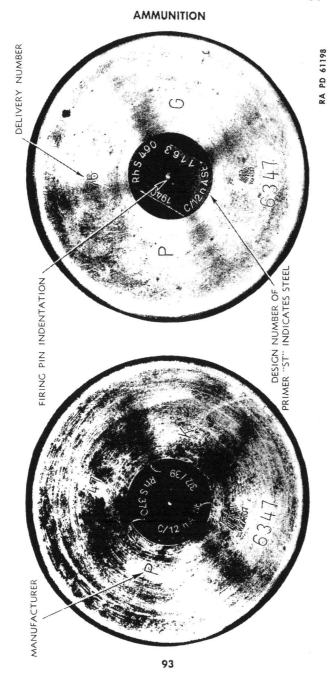

Figure 73 — German 88-mm Cartridge Cases — Base End Views

RA PD 61198

DELIVERY NUMBER

FIRING PIN INDENTATION

DESIGN NUMBER OF
PRIMER "ST" INDICATES STEEL

MANUFACTURER

GERMAN 88-MM ANTIAIRCRAFT GUN MATERIEL

32.74 pounds and is 34.21 inches in length. Muzzle velocity and weight of projectile are given in Table I. Packing of this round is described in paragraph 66.

(2) PROJECTILE. The two rotating bands are bimetallic, being composed of copper electroplated on an iron band. Another type of rotating band may be of ductile iron. The projectile contains a bursting charge of TNT which is approximately 1.8 percent of the total weight of projectile. Weight of the tracer composition is 13 grams. The windshield is attached to the armor-piercing cap by spot welding at 12 places. The projectile has a square base and is 14.49 inches in length. The fuze is described in paragraph 65.

(3) PENETRATION AGAINST HOMOGENEOUS ARMOR PLATE.

TABLE III. PENETRATION DATA OF A.P.C. PROJECTILE AGAINST HOMOGENEOUS PLATE

Range in Yards	Thickness of Plate in Inches	
	Normal Impact	Impact at 30 Degrees
500	5.07	4.33
1,000	4.68	3.97
1,500	4.33	3.62
2,000	3.93	3.30

(a) In addition to Table III above, the armor-piercing ammunition is effective against smaller concrete emplacements, particularly if they have exposed perpendicular walls. Eight well-grouped armor-piercing shells at 800-meter (8.75-yd) range is sufficient to penetrate 2 meters (2.2 yds) of reinforced concrete.

(4) PREPARATION FOR FIRING. This round is ready for firing when removed from its packing.

65. FUZES.

a. General. The point fuzes used with the German 88-mm high-explosive shells consist of the following:

Zt. Z. S./30 .... Time fuze (30-second) with spring-wound action
Zt. Z. S/30 Fg[1] ... Time fuze (30-second) with inertia-operated action
A.Z. 23/28 .. Percussion fuze, superquick or delay (0.11 second) action

(1) These fuzes, and their markings, are shown in figures 74 and 75. With the armor-piercing capped shell, a base-detonating (Bodenzunder, Bd. Z.) fuze is used. It appears that none of these fuzes are boresafe. The diameter over the threads of the point fuzes is 1.96 inch, and the pitch of the threads is 3-mm or 0.12 inch.

CAUTION: Fuzes will not be disassembled. Any attempt to disassemble fuzes in the field is dangerous, and is prohibited except under specific directions from the Chief of Ordnance.

AMMUNITION

INERTIA-OPERATED TIME FUZE

RA PD 61196

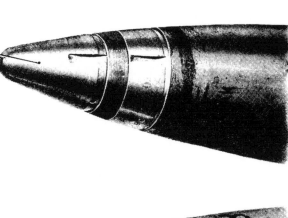

PERCUSSION FUZE

SPRING-WOUND TIME FUZE

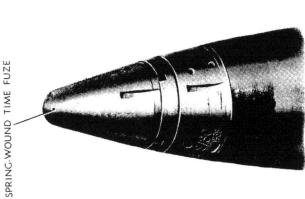

Figure 74 — Spring-wound Time, Percussion, and Inertia-operated Time Fuzes for German 88-mm High-explosive Shell — View Showing Setter Grooves and Selector Element

GERMAN 88-MM ANTIAIRCRAFT GUN MATERIEL

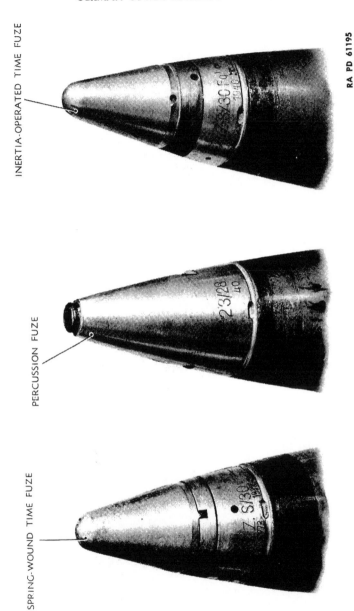

RA PD 61195

INERTIA-OPERATED TIME FUZE

PERCUSSION FUZE

SPRING-WOUND TIME FUZE

Figure 75 — Spring-wound Time, Percussion, and Inertia-operated Time Fuzes for German 88-mm High-explosive Shell — View Showing Fuze Markings

AMMUNITION

b. **German Fuzes, Zt. Z. S/30 and Zt. Z. S/30 Fg¹.**

(1) DESCRIPTION. These fuzes, shown in figures 74 and 75, are 30-second time fuzes. These do not have impact elements. It will be noted that there are no graduations on the time ring. For use, the fuze must be set by means of the fuze setter provided. The zero setting of the fuze is "recess over recess;" on the fuze setter, the indicator (arrow) is at zero. The fuze setter is set at the fuze setting found in the firing table and the fuze is then set as described in chapter 4, section III. A time-safety feature in the fuze prevents time action below 2 seconds of flight. The Zt. Z. S/30 has a spring-wound mechanical time mechanism, whereas the Zt. Z. S/30 Fg¹ has an inertia-operated mechanical time mechanism similar to U. S. FUZE, time, mechanical, M43.

(2) PREPARATION FOR FIRING. The fuzes are prepared for firing as described in subparagraph b (1), above. Fuzes which have been set on rounds prepared for firing but not fired, must be reset at zero. This resetting is accomplished in the same manner as in setting, but with the fuze setter indicator at zero.

c. **German Fuze, A.Z. 23/28.**

(1) DESCRIPTION. This fuze, shown in figures 74 and 75, is similar to the A.Z. 23 used with the German 105-mm howitzer high-explosive shell. The number "28" apparently identifies this fuze for use with German 88-mm shell. The fuze contains two actions, superquick (ohne versögerung, O.V.) and delay (mit versögerung, M.V.). Although both actions are initiated on impact, the functioning of the shell depends upon the setting of the selector of the fuze. Unlike the U.S. FUZE, P.D., M48, it should be noted that there is only one firing pin; should this fail, the projectile will become a dud. However, it appears that the firing assembly is more sensitive to impact than the U.S. FUZE, P.D., M48. Also, unlike the Fuze M48, the German fuze, is not a boresafe fuze. As shipped, the fuze is set for superquick action; that is, the slot on the setting sleeve of the selector is parallel to the axis of the fuze and is thus alined with the registration line marked "O." To set the fuze for delay action, the slotted setting sleeve is turned 90 degrees so that the slot is alined with the line marked "M" on one side of the setting sleeve and with "V" on the other side. The delay action is provided by a delay pellet of 0.11 second delay. The setting may be changed at will with a screwdriver or with "SETTING KEY A.Z. 23" (Stellschlüssel Für A.Z. 23) at any time before firing. This can be done even in the dark by noting the position of the slot. The slot is parallel to the fuze axis for superquick ("O") action, or at right angles thereto for delay action ("M" and "V"). Figure 77 illustrates the internal parts of the fuze.

(2) PREPARATION FOR FIRING. As shipped, the fuze is ready for firing with superquick action. To set the fuze for delay action, the setting sleeve is turned with the setting key or screwdriver, as de-

GERMAN 88-MM ANTIAIRCRAFT GUN MATERIEL

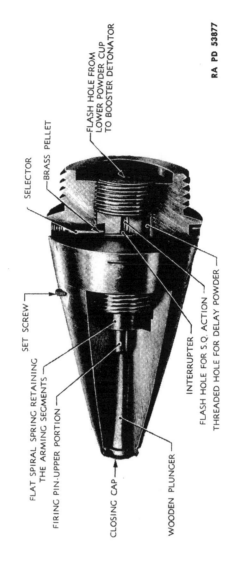

RA PD 53877

FLASH HOLE FROM
LOWER POWDER CUP
TO BOOSTER DETONATOR

SELECTOR

BRASS PELLET

SET SCREW

FLAT SPIRAL SPRING RETAINING
THE ARMING SEGMENTS

FIRING PIN-UPPER PORTION

CLOSING CAP

WOODEN PLUNGER

INTERRUPTER

FLASH HOLE FOR S.Q. ACTION

THREADED HOLE FOR DELAY POWDER

Figure 76 — German A.Z. 23 Fuze — Sectional View

AMMUNITION

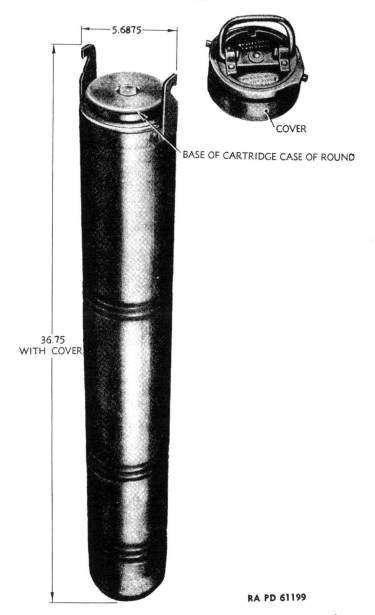

5.6875

COVER

BASE OF CARTRIDGE CASE OF ROUND

36.75
WITH COVER

RA PD 61199

Figure 77 — 1-Round Metal Container for German 88-mm Round

99

GERMAN 88-MM ANTIAIRCRAFT GUN MATERIEL

scribed in step (1), above, through 90 degrees so that the slot on the setting sleeve is alined with the letters "M" and "V." Fuzes which have not been fired should be reset to superquick or "O.V." The slot will then be in line with "O."

d. German Base-detonating Fuze, Bd. Z. This fuze which is assembled in the base of the armor-piercing capped projectile, is a nondelay type. The fuze mechanism is standard for use with other caliber armor-piercing projectiles, such as in the German 7.5 cm., 8.8 cm., and 10.5 cm. rounds. However, the body of the fuze differs depending upon the caliber. The complete fuze weighs 2.18 pounds. The tracer assembly is threaded into the base of the fuze body and the detonator assembly is threaded into the forward portion. The safety feature consists of five brass safety blocks, which are held in the unarmed position by a flat, circular spring, and which engage the shoulder of the primer housing to restrain it against forward movement. Upon rotation of the projectile, centrifugal force causes the safety blocks to move outward against the flat spring, thereby arming the fuze. Upon impact, the primer moves forward, impinging the primer against the firing pin. The resulting primer flame passes into a retaining jet which directs the gases into the detonator initiating the charge. It is believed that a *short delay* is obtained through the plunger action of the primer assembly and also by retardation of the gases by the jet, which prevents the gases from functioning the detonator until considerable pressure has been developed.

66. PACKING.

a. General. German 88-mm rounds are packed in individual sealed steel containers (fig. 77) particularly for use in the tropics or three per wicker basket (fig. 78).

b. Steel Containers. The steel container (fig. 77) is hermetically sealed by a rubber gasket under a removable steel cover. As shipped with one complete round, it weighs approximately 47 pounds and its calculated volume is approximately 2.2 cubic feet. It is painted a slate gray color. Two tags are pasted on the cover.

(1) One tag, of black paper is printed with white ink as follows:
"8.8 cm. Pzgr. Patr.
Bd. Z. f. 8.8 cm. Pzgr."

(2) The other tag of white paper is printed in red as follows:
"Fur Tropen!
Normale Pulvertemperatur
+25°C."

(a) The cover is constructed of several parts arranged so that when the cover is placed in position against the two hooks, and its handle is turned in a clockwise direction, pressure is applied to a rubber gasket between the cover and container body to effect a seal.

AMMUNITION

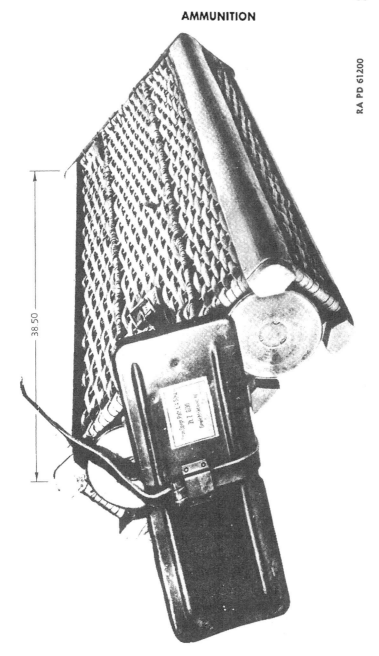

RA PD 61200

Figure 78 — 3-Round Wicker-type Container for German 88-mm Rounds

38.50

GERMAN 88-MM ANTIAIRCRAFT GUN MATERIEL

A hair felt disk is used as a cushion between the head of the cartridge case and the cover.

c. **Wicker-type Container.** The wicker-type container (fig. 78) contains three rounds. A metal cover is held in place by a leather strap. The base is also of metal. Round rubber stops at the bottom of the container protect the fuzes against jarring. The rest of the container is of wood and wicker construction. The dimensions of the container are 38½ by 14¾ by 5½ inches.

67. INTERCHANGEABILITY OF AMMUNITION ITEMS.

a. The British No. 18 primer for 40-mm ammunition is interchangeable with the German primer, C/12nA. In addition, other calibers of German ammunition, such as German 105-mm howitzer ammunition, have the German C/12nA primer as a component of the cartridge case.

68. TROPICAL AMMUNITION.

a. Ammunition for use in the tropics is marked in red lettering, as follows: "P.T. +25°C." This marking appears on the side of cartridge cases. Shell for use in the tropics may be marked "Tp."

b. Containers for tropical ammunition have the following marked in red on white labels:

"Fur Tropen
Normale Pulvertemperatur
+25°C."

c. Tropical ammunition has reduced weight of propellant and gives normal range table performance at +25 C (77 F). Where tropical ammunition has not been issued or manufactured, special range tables are provided for use in the tropics with standard ammunition. The temperature taken as normal for standard ammunition is 10 C (50 F).

69. PRECAUTIONS IN HANDLING CAPTURED AMMUNITION.

a. *All* captured ammunition should be examined by qualified personnel as soon as practicable. Loose ammunition may be dangerous and is rarely worth the trouble of collection.

b. Ammunition may be dangerous because of:
(1) Deliberate "booby traps" laid by the enemy.
(2) Having been subject to fire or shelling.
(3) Removal of safety devices from fuzes, etc. (either deliberate or accidental).
(4) Exposure rendering explosive elements unreliable.

c. Ammunition known or suspected of being dangerous will not be moved or touched, but destroyed in accordance with directions in TM 9-1900.

AMMUNITION

d. Destroyed ammunition should be salvaged for brass parts. In addition, all enemy airtight containers should be returned to the base. This also applies to timber and to wooden boxes suitable for use as dunnage or for remaking ammunition boxes.

e. Ammunition should be recovered by complete rounds; for example, unfuzed shell are useless without the appropriate fuzes.

f. Personnel handling captured ammunition should keep in mind the fact that although two types of ammunition appear to have identical measurements, they are not necessarily interchangeable. Experiments to ascertain interchangeability are forbidden except by special authority.

g. No unauthorized modifications or experimentation will be carried out on any ammunition.

70. CARE, HANDLING, AND PRESERVATION.

a. In addition to the precautions and care in handling U.S. ammunition as given in TM 9-1900, the following apply particularly to German 88-mm ammunition.

(1) The fuze A.Z. 23/28 is particularly sensitive; hence, it is important that the path of flight before the muzzle be free of all obstructions, including small branches and leaves. Otherwise, premature burst may occur.

(2) Components of ammunition prepared for firing but not fired will be returned to their original condition and packing.

(3) Projectiles with impact fuzes (A.Z. fuzes) whose top or forward closing disk has been so damaged that the firing pin is pressed down or has fallen out, will not be fired. They are, however, safe to transport.

(4) Projectiles with time fuzes (Zt. Z. fuzes) may not be fired when the rotatable closing cap of the fuze is bent, dented, or damaged, or cannot be turned by the fuze setter. However, they are safe to transport.

(5) Rounds which have fallen and have not been damaged may be fired.

(6) After each round is fired, it is necessary to examine the bore of the weapon to determine whether any foreign matter remains in the bore. All particles or obstructions should be removed to prevent jamming of the weapon upon firing the next round.

(7) The primer must be hit dead center or it may not function.

71. FIELD REPORT OF ACCIDENTS.

a. When an accident involving the use of ammunition occurs during training practice, the procedure prescribed in section VII, AR 750-10, will be observed by the ordnance officer under whose supervision the ammunition is maintained or issued. Where practicable, reports covering malfunctions of ammunition in combat will be

GERMAN 88-MM ANTIAIRCRAFT GUN MATERIEL

made to the Chief of Ordnance, giving the type of malfunction, type
of ammunition, the lot number of the complete rounds or separate-
loading components, and condition under which fired.

**72. GERMAN ABBREVIATIONS AND TERMINOLOGY OF
AMMUNITION ITEMS.**

a. General. The following abbreviations, symbols, and terms may
be found on labels or in communications and literature pertaining to
the ammunition items described herein. Certain general terms are also
included.

b. Abbreviations.

TABLE V. GERMAN AMMUNITION ABBREVIATIONS

A.Z.	Aufschlagzünder	percussion fuze
A.Z. m. V.	Aufschlagzünder mit versögerung	percussion fuze with delay action
Bd. Z.	Bodenzünder	base percussion fuze
Bl.	Blindgänger	dud
Bl. P.	Blättchenpulver	flaked gunpowder
Bz.	Brennzünder	time fuze (powder train type)
Digl. Dgl.	Diglycol	diglycol
Dopp. Z. D.Z.	Doppelzünder	combination fuze
Ex. Mun.	Exerziermunition	dummy ammunition; blank ammunition
f	Für	for
Flak.	Flugabwehrkanone	antiaircraft gun
Flb.	Flugbahn	trajectory
Fp.	Füllpulver	high explosive
Gesch	Geschoss	projectile; shell
G. Gr.	Gasgranate	gas shell
Gr.	Granate	shell
Grf.	Granatfüllung	bursting charge of shell
Gr. m. p.	Granate mit Panzerkopf	armor-piercing shell
Gr. Z. G.Z.	Granatzünder	shell fuze
H	Hexagen	cyclonite, R.D.X.
Kl.	Klein	small
Kp	Krupp	Krupp
Kz.	Kopfzünder	point-detonating fuze

TABLE V. GERMAN AMMUNITION ABBREVIATIONS (Cont'd)

Ldg. L.	Ladung	charge; propelling charge; load
Lggr.	Langgranate	long shell
Lv.	Ladungsverhältnis	ratio of charge to weight of projectile
m.	mit	with
Mun.	Munition	ammunition
m.v.	mit versögerung	with delay (fuzes)
Nb.	Nebel	smoke
Nbgr.	Nebelgranate	smoke shell
Ngl. Nigl.	Nitroglyzerin	nitroglycerin
Np.	Nitropenta	P.E.T.N.; penthrite
Nr.	Nummer	number
o.	ohne	without
o.v.	ohne versögerung	without delay (superquick)
Pak.	Panzerabwehrkanone	antitank gun
P. K.	Pulverkasten	ammunition box
P. S. Gr.	Panzerstahlgranate	steel armor-piercing shell
P. T.	Pulvertemperatur	ammunition temperature
Pzgr. Pz. Gr.	Panzergranate	armor-piercing shell
Pz. Spr. Gr.	Panzersprenggranate	high-explosive armor-piercing shell
Sch. Tf.	Schusstafel	firing table
Sch. Z. Schr.	Schlagzündschraube	threaded base percussion fuze
Sonderkart	Sonderkartusche	special charge
Sprgr. Spr. Gr.	Sprenggranate	high-explosive shell
St.	Stahl	steel
Tp.	Tropen	Tropics
Ub.	Ubung	practice
Ubgr. Ub. Gr.	Ubungsgranate	practice shell
v.	versögerung	delay (fuzes)
Z.	Zünder	fuze
Zdschr.	Zündschraube	threaded percussion primer
Zt. Z. ZZ.	Zeitzünder	time fuze

GERMAN 88-MM ANTIAIRCRAFT GUN MATERIEL

c. Glossary.

TABLE VI. GERMAN AMMUNITION TERMS

Aufschlagzünder (A.Z.) ....percussion fuze	Hexagen (H.) ....cyclonite, R.D.X.
Aufschlagzünder mit versögerung ....percussion fuze with delay action	Holzkasten ....wooden box
	Hülse ....cartridge case
	Hülsenbezeichnung ....cartridge case designation (number)
Blättchenpulver (Bl.P.) ....flaked gunpowder	Kartusche ....cartridge case
Bleidraht ....lead wire	Kartuschhülse ....cartridge case
Blindgänger (Bl.)........dud	Kartuschkorb ....ammunition basket
Bodenzünder (Bd. Z.) ....base percussion fuze	Kartuschvorlage ....cartridge case wad; flash reducer
Brennzünder (Bz.)......time fuze (powder train type)	Kennbuchstabe ....identification mark
Brisanz ....high-explosive	Klein (Kl.) ....small
Brisanzgeschoss ....high-explosive shell.	Kopfzünder (Kz.) ...point-detonating fuze
Brisanzmunition ....high-explosive ammunition	Ladung (Ldg.; L.).....charge; propelling charge; load
Diglycol (Digl.; Dgl.). diglycol	
Doppelzünder (Dopp. Z.; D.Z.)....combination fuze	Ladungsverhältnis (Lv.) ....ratio of charge to weight of projectile
Exerziermunition (Ex. Mun.) ....dummy ammunition; blank ammunition	Langgranate (lggr.)....long shell
	Lieferungsnummer ....delivery number
Flugabwehrkanone (Flak) ....antiaircraft gun	mit (m.) ....with
	mit versögerung (m.v.) ....with delay (fuzes)
Flugbahn (Flb.) ....trajectory	
Füllpulver (Fp) ....high explosive	Munition (Mun.)........ammunition
Für (f) ....for	Nebel (Nb.) ....smoke
Gasgranate (G. G.)....gas shell	Nebelgranate (Nbgr.) ....smoke shell
Geschoss (Gesch.)......projectile; shell	
Gewichtsklasse ....weight class shell	Nebelgeschoss ....smoke shell
Granate (Gr.) ....shell	Nitroglyzerin (Nigl.; Ngl.) ....nitroglycerin
Granate mit Panzerkopf (Gr. m. P.)......armor-piercing shell	Nitropenta (Np.) ......P.E.T.N.; penthrite
Granatfüllung (Grf.)..bursting charge of shell	Nummer (Nr.) ....number
	ohne (o.) ....without
Granatzünder (Gr. Z.; G. Z.)........shell fuze	ohne .versögerung (o.v.) ....without (superquick)
Haube ....ballistic cap	
Hauptladung ....propellant (lit; main charge)	

AMMUNITION

TABLE VI.   GERMAN AMMUNITION TERMS (Cont'd)

Panzerabwehrkanone (Pak.) ....antitank gun	Schlagzündschraube (Schl. Z. Schr.)......threaded base percussion fuze
Panzergranate (Pzgr.; Pz. Gr.)......armor-piercing shell	Sonderkartusche (Sonderkart.) ....special charge
Panzersprenggranate (Pz. Spr. Gr.)........high explosive armor-piercing shell	Sprenggranate (Sprgr.; Spr. Gr.)...high-explosive shell
Panzerstahlgranate (P. S. Gr.)...............steel armor-piercing shell	Stahl (St.) ....steel
Pulver .... powder	Stellschlüssel ....setting key (fuzes); hand fuze setter; adjusting wrench
Pulverkasten (P.KK.) .... ammunition box	Tropen (Tp.) ....Tropics
Pulverladung ....powder charge	Ubung (Ub.) ....practice
Pulvertemperatur (P. T.) ....powder temperature	Ubungranate (Ubgr.; Ub. Gr.)...practice shell
rauchloses Pulver.......smokeless powder	Versögerung (V.)........delay
rauch-schwaches Pulver ....smokeless powder	Vorlage ....flash hider
rohrsicherer Zünder....bore-safe fuze	Zünderstellung ....fuze setting
Schusstafel (Sch. Tf.) ....firing table	Zünderschlüssel ....hand fuze setter
	Zünderstellmaschine ..fuze setter
	Zünderstellschlüssel ..hand fuze setter
	Zündschraube ....threaded percussion primer
	Zeitzünder (Zt. Z.; ZZ.)............time fuze

d. Index Numbers on German Shell Indicating Type of High-explosive Filler.

TABLE VII.   NUMBERS ON GERMAN SHELL INDICATING TYPE OF H. E. FILLER

No. on Shell	Type of Filler
1	...Fp 02 (TNT) in paper or cardboard container
2	...Grf 88 (picric acid) in paper or cardboard container
10	...Fp 02 plus Fp 5 plus Fp 10 (TNT fillers) in paper or cardboard container
13	...Fp 40/60 (40-60 amatol, poured)
14	...Fp 02 (TNT), poured
32	...Np 10 (P. E. T. N. filler) in paraffin-waxed paper wrapping
36/38	...Np 40 plus Np 60 (P. E. T. N. fillers) in paraffin-waxed paper wrapping
91	...H 5 (Cyclonite; R.D.X.) in paraffin-waxed paper wrapping

GERMAN 88-MM ANTIAIRCRAFT GUN MATERIEL

e. German Explosives, Abbreviations.

TABLE VIII. GERMAN EXPLOSIVES, ABBREVIATIONS

Abbreviation	German Nomenclature	English Equivalent
Fp 02	Füllpulver 02	TNT
Fp 5	Füllpulver 5	TNT with 5 percent montan wax
Fp 10	Füllpulver 10	TNT with 10 percent montan wax
Fp 40/60	Füllpulver 40/60	40-60 amatol, poured
G of 88	Granatfülling 88	picric acid
H	Hexagen	cyclonite, R.D.X.
H5	Hexagen 5	cyclonite with 5 percent montan wax
Np	Nitropenta	P.E.T.N.; penthrite
Np 10	Nitropenta 10	P.E.T.N. with 10 percent montan wax
Np 40	Nitropenta 40	P.E.T.N. with 40 percent montan wax
Np 65	Nitropenta 65	P.E.T.N. with 65 percent montan wax

CHAPTER 4

SIGHTING AND FIRE CONTROL EQUIPMENT

Section I

INTRODUCTION

73. INTRODUCTION.

a. The sighting and fire control equipment for the 8.8 cm Flak 18 or 36 varies, depending on the use to which the weapon is put. The gun may be used for direct fire as in antitank work, for indirect fire, or for antiaircraft fire.

b. For direct fire the telescopic sight ZF.20E (Zielfernrohr 20E) is used for laying the gun in azimuth and elevation. This sight, consisting of elbow telescope, telescope mount, and range drum is mounted on a bracket geared to the elevation quadrant on the right-hand size of the top carriage. The gun is laid in elevation by matching its pointer on the elevation quadrant against a pointer controlled by the telescopic sight.

c. The gun may be laid in azimuth and elevation against aircraft, moving ground or sea targets in accordance with data obtained from a director. Either of two directors may be used. The stereoscopic director Kdo. Gr. 36 (Kommandogerät 36) connects electrically to the on-carriage equipment for blackout dial matching. The auxiliary director Kdo. Hi. Gr. 35 (Kommandohilfsgerät 35) develops data which are telephoned to the gun crew. It is much lighter than the Kdo. Gr. 36 and is intended for mobile use. It is designed to be carried by porter bar.

d. When the stereoscopic director Kdo. Gr. 36 is used, the off-carriage equipment includes a distribution box; a switchboard with battery source of power for telephone and transmission of data; cables connecting the director to the distribution box, switchboard, and guns. The on-carriage equipment includes the instruments listed below:

(1) The panoramic telescope Rbl. F. 32 (Rundblickfernrohr 32) which is placed in a telescope holder on the top of the recuperator for initial orientation of the guns with the director.

(2) The azimuth indicator.

(3) The elevation indicator.

(4) The fuze setter.

(5) On-carriage wiring and boxes.

e. When the auxiliary director Kdo. Hi. Gr. 35 is used, a 4-meter range finder (separately connected) furnishes slant range.

f. Information is not available on the switchboard, the off-carriage cables, or the height finder telescopes of the Kdo. Gr. 36.

GERMAN 88-MM ANTIAIRCRAFT GUN MATERIEL

CHAPTER 4

SIGHTING AND FIRE CONTROL EQUIPMENT (Cont'd)

Section II

SIGHTING EQUIPMENT

74. TELESCOPIC SIGHTS (ZF.20E OR ZF.20) WITH ELEVA-TION QUADRANT FOR DIRECT FIRE.

a. **Description of Telescopic Sight (ZF.20E).**

(1) The telescopic sight Flak ZF.20E (ZF, Zielfernrohr, meaning telescopic sight) (figs. 79 to 81) is used for direct fire. It includes an elbow telescope, angle of site mechanism, range quadrant and elevation mechanism, and deflection mechanism.

(2) ELBOW TELESCOPE. The elbow telescope is of fixed focus type, has 4X magnification and a field of view of 17 degrees 30 minutes. The image remains erect during the rotation of the telescope. The filter knob and detent positions each of the filters: clear, green, light neutral and very dark neutral. A window and a dovetail surface permit attachment of a lamp for illumination of the reticle. The reticle pattern (fig. 81) permits rapid centering of reticle on target in desert glare and has indications for estimating lead.

(3) ANGLE OF SITE MECHANISM. The angle of site mechanism rotates the telescope and deflection mechanism within the main housing. The angle of site scale is graduated in 100-mil intervals from plus to minus 200 mils. Zero is normal. The angle of site micrometer is graduated in 1-mil intervals from 0 to 100 mils. Stop rings limit motion within the limits of the scale.

(4) RANGE QUADRANT AND ELEVATION MECHANISM. The range and elevation knob rotates the entire telescope and angle of site mechanism in the housing. On each end of the worm is a drum and pointer. One drum indicates elevation in degrees and is graduated from 0 to 12 degrees in $\frac{1}{16}$-degree intervals. The other drum is graduated in 100-meter intervals from 0 to 94 meters.

(5) DEFLECTION MECHANISM. Deflection may be set into the sight by rotation of the deflection knob. The deflection scale is graduated from 250 mils left to 250 mils right.

(6) A blank eyeshield is provided to rest the observer's eye. The shield may be adjusted for interpupillary distance.

(7) The telescopic sight has two trunnions and a clamp for installation on the telescopic sight bracket of the mount.

Figure 79 — Telescopic Sight ZF.20E — Front View

RA PD 70160

TRUNNION

OBJECTIVE

CLAMP

RANGE DRUM

EYEPIECE

ELEVATION MICROMETER

BLANK EYESHIELD

RETICLE WINDOW

OPEN SIGHT

FILTER KNOB

AZIMUTH DEFLECTION SCALE

Figure 80 — Telescopic Sight ZF.20E — Top View

RA PD 70161

SIGHTING EQUIPMENT

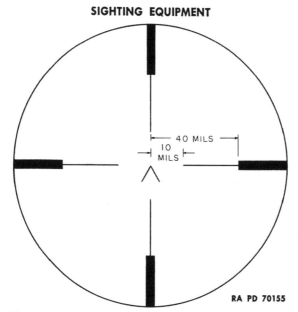

RA PD 70155

Figure 81 — Telescopic Sight ZF.20E — Reticle Pattern

b. Description of Telescopic Sight (ZF.20). The telescopic sight (ZF.20) (figs. 82, 83, and 84) is similar to the telescopic sight (ZF.20E) but lacks the range drum. The gun commander must memorize the range equivalent to some of the elevation scale readings.

c. Elevation Quadrant.

(1) The elevation quadrant (fig. 85) is used in direct fire sight (ZF.20E or ZF.20) for laying the gun in elevation. The elevation quadrant includes a quadrant, an outer pointer, an inner pointer, and a link to the telescopic sight bracket.

(2) The quadrant is centered on the cradle trunnion and is fastened to the mount. In operation the quadrant remains stationary. The quadrant is graduated in 0.25-degree intervals from minus 3 degrees in depression to plus 85 degrees in elevation.

(3) The outer pointer (quadrant elevation pointer) is fastened to the cradle trunnion. It bears an index which alines with the quadrant scale, indicating the elevation of the gun.

(4) The inner pointer (direct sight elevation pointer) pivots on the cradle trunnion and moves with the bracket of the telescopic sight. The pointer has an index line which alines with the index of the outer pointer.

(5) The link connects the inner pointer with the bracket of the telescopic sight and has a turnbuckle for alinement of the sight with the gun bore.

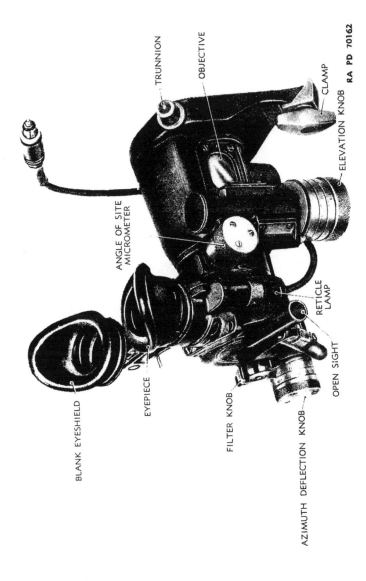

TRUNNION

OBJECTIVE

CLAMP

ELEVATION KNOB

RA PD 70162

ANGLE OF SITE MICROMETER

RETICLE LAMP

BLANK EYESHIELD

EYEPIECE

OPEN SIGHT

FILTER KNOB

AZIMUTH DEFLECTION KNOB

Figure 82 — Telescopic Sight ZF.20 — Front View with Eyeshields Swung Down and Forward

114

SIGHTING EQUIPMENT

RA PD 70163

Figure 83 — Telescopic Sight ZF.20 — View of Bottom from Front

GERMAN 88-MM ANTIAIRCRAFT GUN MATERIEL

RA PD 70164

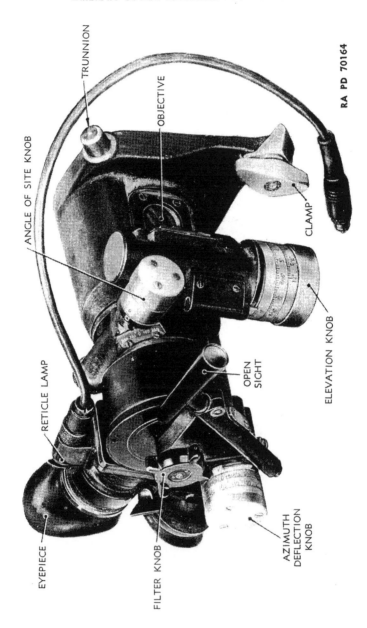

TRUNNION

OBJECTIVE

ANGLE OF SITE KNOB

CLAMP

RETICLE LAMP

ELEVATION KNOB

OPEN SIGHT

EYEPIECE

FILTER KNOB

AZIMUTH DEFLECTION KNOB

Figure 84 — Telescopic Sight ZF.20 — Side View

SIGHTING EQUIPMENT

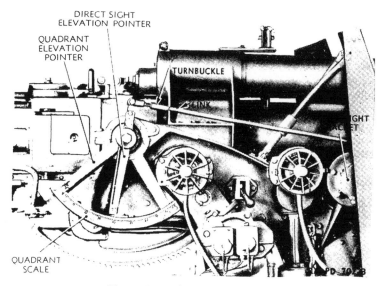

Figure 85 — Elevation Quadrant

(6) When the elevation quadrant, the link, and the telescopic sight are in adjustment, the gun may be laid in elevation by setting the desired elevation or range in on the telescopic sight and matching the pointers on the elevation quadrant.

d. Operation.

(1) INSTALLATION. Examine trunnions and mating surfaces for nicks and burs. Set the trunnions of the telescopic sight in the proper slots of the telescope bracket and clamp the telescopic sight firmly in place.

(2) BORE SIGHTING.

(a) Remove the firing pin and use the firing pin hole as the breech bore sight. Improvise the muzzle bore sight by fastening two pieces of cord or wire across the muzzle face, locating the cord in the horizontal and vertical grooves on the muzzle face and securing with a strap around the barrel close to the muzzle.

(b) Lay the gun on a well defined distant aiming point by sighting through the bore. Turn the handwheel of the telescopic sight bracket (just below the bracket) to match the direct sight elevation pointer with the quadrant elevation pointer. Set the range scale, the angle of site scale, and the deflection scale of the telescopic sight to "0." The telescope sight should now be on the distant aiming point.

(c) If error in elevation exists, lengthen or shorten the link bar by adjustment of the turnbuckle.

GERMAN 88-MM ANTIAIRCRAFT GUN MATERIEL

(d) Errors in azimuth may be corrected by turning the telescope deflection micrometer until the telescope alines in azimuth with the distant aiming point, then loosening the three screws in the end of the micrometer, slipping the scale to zero, and reclamping.

(3) DIRECT FIRE OPERATION.

(a) The azimuth gun pointer sits behind the telescopic sight with the traversing handwheel at his left.

(b) The angle of site scale and micrometer should be set at zero.

(c) The elevation for range in degrees and $\frac{1}{16}$ degrees is set in on the elevation scale or in meters (on ZF.20E only) on the range drum.

(d) The deflection in mils, if large, is set in on the deflection drum. (Red graduations bring the piece to the left.)

(e) Small deflections may be estimated on the reticle (fig. 81).

(f) The gun is laid for elevation by matching both pointers of the elevation quadrant.

(g) The azimuth gun pointer tracks the target with the traversing handwheel.

CHAPTER 4

SIGHTING AND FIRE CONTROL EQUIPMENT (Cont'd)

Section III

FIRE CONTROL EQUIPMENT

75. COMPONENTS.

a. The fire control equipment consists of the stereoscopic director 36 (Kommandogerät 36), or the auxiliary director 35 (Kommandohilfsgerät 35); the azimuth and elevation indicators; the fuze setter; the necessary on-carriage wiring; and the cables.

b. When the stereoscopic director is available, it transmits data electrically to the indicators and fuze setter. These have dials, each containing 3 concentric circular rows of 10 lamps each. The data are indicated by the lighting of lamps and matched by operating the gun until the 3 blackout indexes of each of the 2 indicators and the fuze setter cover the lighted lamps. The gun is then alined on the target indicated by the director.

c. When the auxiliary director is used, slant range is obtained from a separate 4-meter range finder (information not available) and is manually set into the director. The computed values are read off on dials and scales visible in the sides of the director and then telephoned to the gun crew.

d. Information is not available on the cables used to connect the directors with the guns.

76. PANORAMIC TELESCOPE 32 (RBL. F. 32).

a. Description of Panoramic Telescope 32 (Rbl. F. 32).

(1) The panoramic telescope used for orienting with the stereoscopic director is the Rundblickfernrohr 32 (figs. 86 and 87). The

119

GERMAN 88-MM ANTIAIRCRAFT GUN MATERIEL

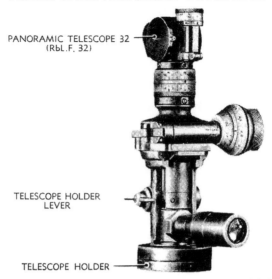

PANORAMIC TELESCOPE 32
(RbL.F. 32)

TELESCOPE HOLDER
LEVER

TELESCOPE HOLDER

RA PD 70165

Figure 86 — Panoramic Telescope 32 (Rbl. F. 32) in Telescope Holder

telescope is clamped in the telescope holder on the top of the recuperator. The holder is located at the pintle center of the mount.

(2) The panoramic telescope (fig. 88) is a 4-power, fixed focus
type with a field of view of 9 degrees. The line of sight may be raised
or lowered by rotation of the angle of site knob. The angle of site
scale is graduated from 100 mils to 500 mils (300 mils is normal).
The angle of site micrometer is graduated in mils from 0 to 100 mils.
The azimuth scales on the vertical barrel of the telescope are graduated in 100-mil intervals; the upper scale, 0 to 64; the lower scale, 0 to
32, 0 to 32. A knurled portion permits adjustment. The center index
is locked in place by a lug at the front of the telescope. The azimuth
micrometer includes 2 scales graduated in mils from 0 to 100 mils.
The index between the scales is fixed. It is believed that one scale is
used for setting in corrections and the other for setting in fine azimuth
values. A throw-out lever is provided for rapid setting in azimuth.
A locking lever locks the azimuth micrometer in any setting. The
reticle pattern is shown in figure 87.

(3) The telescope holder is of conventional design. It is welded
to the top of the recuperator and is in the vertical plane through
the axis of bore. It is also at the pintle center. The telescope holder
lever operates a cam which engages a notch in the body of the telescope. Two screws with jam nuts permit adjustment in azimuth of
the telescope in the holder.

FIRE CONTROL EQUIPMENT

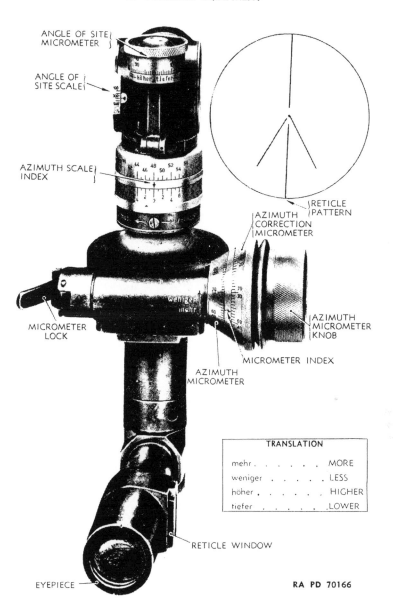

ANGLE OF SITE
MICROMETER

ANGLE OF
SITE SCALE

AZIMUTH SCALE
INDEX

RETICLE
PATTERN

AZIMUTH
CORRECTION
MICROMETER

MICROMETER
LOCK

AZIMUTH
MICROMETER
KNOB

MICROMETER INDEX

AZIMUTH
MICROMETER

RETICLE WINDOW

EYEPIECE

RA PD 70166

TRANSLATION	
mehr	MORE
weniger	LESS
höher	HIGHER
tiefer	LOWER

Figure 87 — Panoramic Telescope 32 (Rbl. F. 32) - Rear View

GERMAN 88-MM ANTIAIRCRAFT GUN MATERIEL

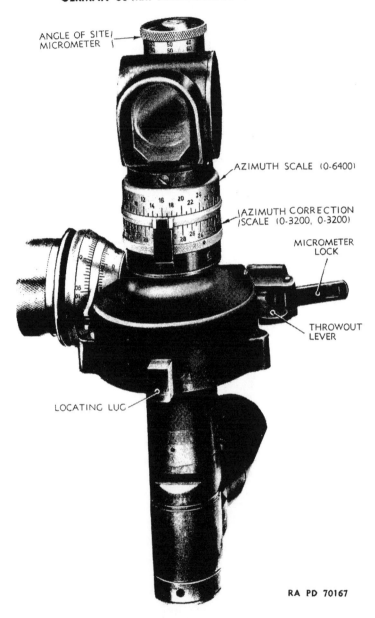

ANGLE OF SITE/
MICROMETER

AZIMUTH SCALE (0-6400)

AZIMUTH CORRECTION
SCALE (0-3200, 0-3200)

MICROMETER
LOCK

THROWOUT
LEVER

LOCATING LUG

RA PD 70167

Figure 88 — Panoramic Telescope 32 (Rbl. F. 32) — Front View

FIRE CONTROL EQUIPMENT

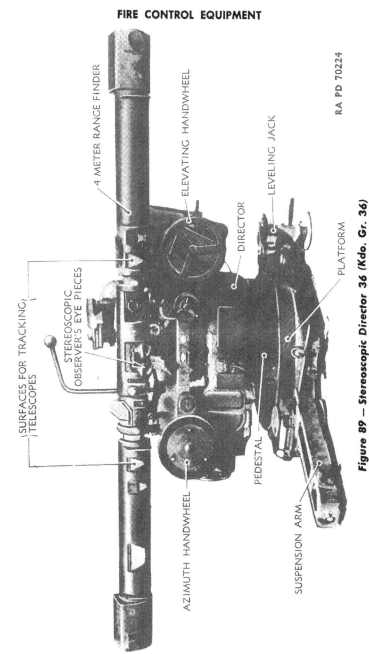

4 METER RANGE FINDER

ELEVATING HANDWHEEL

LEVELING JACK

DIRECTOR

RA PD 70224

PLATFORM

SURFACES FOR TRACKING TELESCOPES

STEREOSCOPIC OBSERVER'S EYE PIECES

AZIMUTH HANDWHEEL

PEDESTAL

SUSPENSION ARM

Figure 89 — Stereoscopic Director 36 (Kdo. Gr. 36)

GERMAN 88-MM ANTIAIRCRAFT GUN MATERIEL

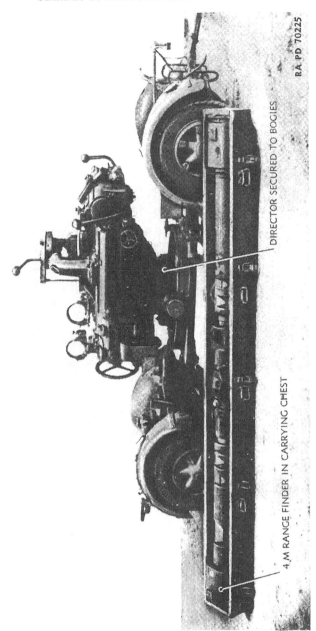

RA PD 70225

DIRECTOR SECURED TO BOGIES

4.M RANGE FINDER IN CARRYING CHEST

Figure 90 — Stereoscopic Director 36 (Kdo. Gr. 36) — Range Finder Removed from Director

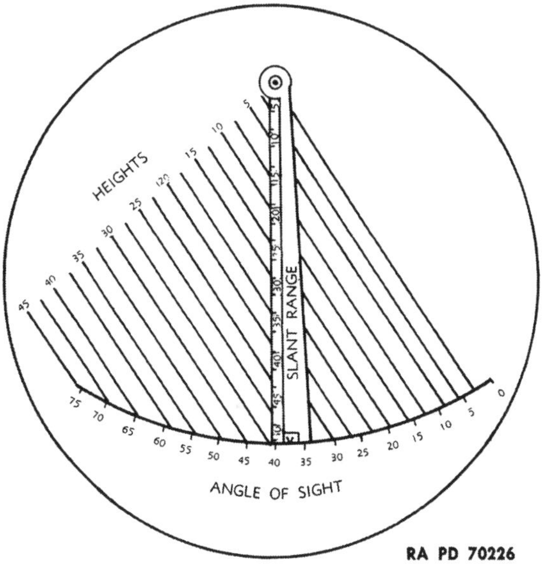

FIRE CONTROL EQUIPMENT

RA PD 70226

Figure 91 — Stereoscopic Director 36 (Kdo. Gr. 36) — Height Scale

77. STEREOSCOPIC DIRECTOR 36 (KDO. GR. 36).

a. General. The stereoscopic director 36 (Kommandogerät 36) (figs. 80 to 86) is the standard director used with the 8.8 cm. flak 36. It is a combined stereoscopic range finder and director, supported on a pedestal which has three leveling feet and two suspension arms for securing to front and rear bogies for travel. The bogies are similar to those used for carrying the gun mount.

b. Range Finder (Em. 4m. R. (H)).

(1) The 4-meter base stereoscopic range finder (Raumbildentfernungsmesser (Höhe)) adapted for height finding has magnification of 12x and 24x and a range scale reading from 500 meters (550 yd) to 50,000 meters (55,000 yd). It is clamped by two rings to the director for use and in travel is carried in a chest fitted with hand grips.

(2) A device for obtaining approximate height is fitted on the right end of the range finder tube. This consists of an arm pivoted at one end and engraved with a scale of ranges. A series of parallel lines graduated to the value of height are engraved on the disk to which the arm is pivoted. As the instrument is elevated, the disk carrying the scale rotates with the instrument. The arm, however, remains vertical, and the height line corresponding to the range on the range arm indicates target height (fig. 91).

125

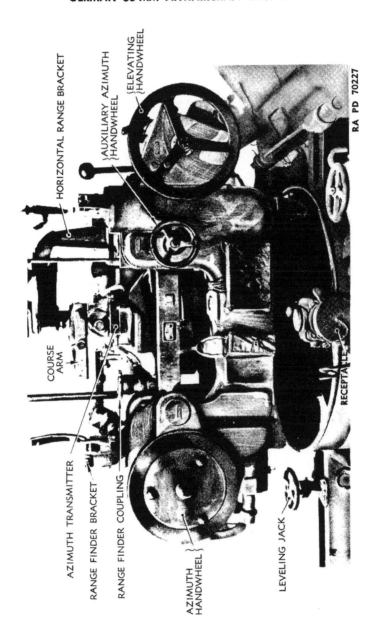

Figure 92 — Stereoscopic Director 36 (Kdo. Gr. 36) — View of Traversing and Elevating Handwheel

RA PD 70227

FIRE CONTROL EQUIPMENT

HORIZONTAL RANGE BRACKET

COURSE ARM

{WIND
{MECHANISM

DA AND RF ARM

RANGE DRUM

{RANGE
{HANDWHEEL

RA PD 70228

HANDWHEEL MATCHING OUTER
POINTER (APPLYING Dv)

HANDWHEEL MATCHING INNER}
POINTER (MEASURING Wv)}

ELEVATION RATE}
HANDWHEEL}

ELEVATING
HANDWHEEL }

LEVELING JACK

Figure 93 — Stereoscopic Director 36 (Kdo. Gr. 36) — View of Range Drum

127

GERMAN 88-MM ANTIAIRCRAFT GUN MATERIEL

(3)   Two tracking telescopes are fitted on the range finder tube to the right and left of the stereoscopic eyepiece. These are employed for keeping the range finder on the target and have a cross wire reticle pattern.

(4)   The range finder is turned in elevation by rotation of the elevating handwheel which is connected by gearing to a gear between the range finder tube and the left bearing ring.

(5)   An optical lath is provided. When it is suspended on the supporting brackets, it provides an artificial infinity for test and adjustment of the range finder.

c. **Director.**

(1)   The director includes a main casting, supported on a pedestal, and supporting a number of box enclosed mechanisms (figs. 92-95). The director determines and transmits the following data to the battery.

(a)   Quadrant elevation.

(b)   Future azimuth.

(c)   Time of flight of projectile, expressed in fuze units.

(2)   The values set into the director are:

(a)   Present angular height.

(b)   Present azimuth.

(c)   Present slant range.

d. **Setting Up.**

(1)   Uncouple the carriage supporting the director from its towing vehicle and lower the platform from the front and rear bogies (fig. 90). Open the cases containing parts and accessories.

(2)   Open the range finder clamp rings of the director. Remove the range finder from its case and clamp it to the director. Couple the range finder to the director. Place the two tracking telescopes and the four other telescopes on their respective mounting surfaces on the range finder. Check against a distant aiming point to make sure that the telescopes are alined with the range finder.

(3)   Level the director by adjusting the three leveling jacks (fig. 89). Level the range finder using the range finder leveling screw and lock nut until the bubble of the level is centered.

(4)   Turn the handwheel for angular height to the left until the zero reading of the angular height graduation of the horizontal range to datum point drum is exactly under the horizontal range to datum point index. Then turn the upper part of the coupling which transfers present angular height to the director until the zero-degree reading of present angular height is visible in the aperture of the present angular height instrument attached to the range finder.

(5)   Remove the range finder eyepiece protector and hang it under the range finder. Set the head rest in the correct position. Hang the arm strap on the knob to the right of the eyepiece. Set the interpupillary distance and focus the eyepieces. If necessary, use the filters.

FIRE CONTROL EQUIPMENT

(6) Bring the panoramic telescope, which is on the top of the recuperator of the gun, to bear on the director. Set this value on the telescope of the director and bodily slew the director to lay on the red stem of the panoramic telescope on the gun.

(7) With the director alined on the gun telescope, set the graduated lateral deflection ring at zero. Set the graduated deflection ring of the course plate at zero.

(8) Screw the counterweights into the range finder and remove the cover from the correcting apparatus. If necessary, install the telescope extension tubes (rain protection) on the range finder.

(9) Adjust the range finder stereoscopically and check alinement of tracking telescopes and observing telescopes and binoculars with the range finder. Detailed information on this is not available at this time.

(10) Wind up the clockwork of the horizontal speed indicator.

(11) Connect the cables between the director, the switchboard, the distribution box, and the guns. Plug in the head and chest set telephones. Detailed information on this is not available at this time.

(12) Set in average values for future azimuth, Q.E., and fuze. Turn on the switch in the receptacle box at the director. Turn on the power with the transmission switch at the switchboard. Check the lamps in the indicators and fuze setter, noting broken or weak lamps. Check the future azimuth, Q.E., and fuze values obtained at the guns with that set in at the director. The director is now ready for operation.

e. **Stereoscopic Director Operation.**

(1) FUNCTIONS OF OPERATORS. It is believed that 11 operators are required.

(a) *Elevation Tracker.* Keeps reticle of elevation tracking telescope (mounted on right range finder tube) laid on the target by turning the elevating handwheel.

(b) *Azimuth Tracker.* Keeps reticle of azimuth tracking telescope (mounted on left range finder tube) laid on the target by turning the azimuth handwheel. An auxiliary elevation handwheel close to the azimuth handwheel enables the azimuth tracker to track in both azimuth and elevation when necessary.

(c) *Stereoscopic Observer.* Tracks the target stereoscopically, indicating slant range on range drum.

(d) *Range or Height Reader.* Reads off elevation figures on the range drum of the range finder to both the horizontal range operator and the target course and speed operator. When range only is being taken, the reader gives these to the horizontal range reader only.

(e) *Horizontal Range Operator.* By operation of the horizontal range handwheel keeps the center circle of the cursor (attached to horizontal range bracket) on the curve indicated on the range drum by the stereoscopic observer.

GERMAN 88-MM ANTIAIRCRAFT GUN MATERIEL

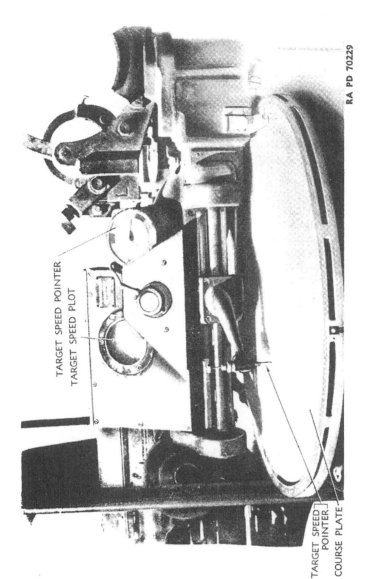

RA PD 70229

TARGET SPEED POINTER
TARGET SPEED PLOT

TARGET SPEED
POINTER
COURSE PLATE

Figure 94 — Stereoscopic Director 36 (Kdo. Gr. 36) — View of Course Plate

RA PD 70230

Figure 95 — Stereoscopic Director 36 (Kdo. Gr. 36) — View of Wind Correction Knobs

GERMAN 88-MM ANTIAIRCRAFT GUN MATERIEL

(f) Target Course and Speed Operator. Watches the trace made by the course pointer over the course plate (fig. 94) and rotates the course bearing disk to keep the parallel lines in agreement with the target course. Keeps the pointer of the horizontal speed scale in agreement with the speed registered on the horizontal speed indicator, or watches the dial of the horizontal speed mechanism, and keeps the colored disk rotating at the same speed as the pointer and the parallel lines of the speed plot in agreement with the track of the plot by turning the handwheel below the course plate.

(g) Present Angular Height Operator. Takes readings from present angular height scale on range drum and sets these in on the present angular height scale of the rate multiplying mechanism. This is done either by direct rotation of the handwheel or by clutching in the variable speed drive and turning the handwheel to control the rate of change. Watches the elevation speed indicator, keeping the inner index matched to the pointer by one handwheel and the outer index matched to the main index by the second handwheel.

(h) Operator for DA and Rf Arm. Matches pointer of DA and Rf arm against pointer moving along course arm by turning the two handwheels mounted on the DA and Rf arm. One handwheel traverses the arm and adds deflection in azimuth to the present azimuth set into the future azimuth transmitter.

(i) Fuze Drum Operator. Reads the future angular height scale (to the right of the drum) and sets the curve indicated on this scale under the center of the cursor by turning the fuze handwheel. This also sets the fuze transmitter.

(j) Quadrant Elevation Operator. Reads future angular height scale (to the left of the drum) and sets the curve indicated on the scale under the center of the cursor by turning the tangent elevation handwheel. This also sets the quadrant elevation transmitter.

(k) Correction Operator. Reads off from the respective registers the drift adjustment, elevation corrections, and fuze setting corrections as shown under the cross threads and sets these in on the respective dials (fig. 95).

f. Operation by Telephone. When transmission of data from the stereoscopic director is desired by telephone, three additional operators are required to telephone the values (Q.E., future azimuth, fuze) to the gun crew.

g. Standby Settings. When firing has ceased, the following settings should be made:
(1) Traverse the director to zero.
(2) Bring the present angular height reading to zero.
(3) Set target height in range finder to 2,500 meters.
(4) Set horizontal range to 5,000 meters.
(5) Set horizontal velocity to 50 meters per second and bring the pointer under its arm.

FIRE CONTROL EQUIPMENT

(6) Set correction for drift to zero.

(7) Set future horizontal range to 5,000 meters.

(8) Set gun elevation to 60 degrees.

(9) Shut off power.

78. AUXILIARY DIRECTOR 35 (KDO. HI. GR. 35).

a. Description.

(1) The auxiliary director 35 (Kommandohilfsgerät 35) (figs. 96 to 99) is a portable director, smaller and less complicated than the Kdo. Gr. 36. Data computed in the director is telephoned to members of the gun crew, who then set in elevation, azimuth, and fuze time. The slant height or range scales are calibrated for use with the 8.8 cm Flak 18 and 10.5 cm Flak 38. Provision is made for two tracking telescopes although information is not available at this time on these.

(2) The slant range prediction is approximated by adding range rate times future time of flight to the present slant range which is obtained from a 4-meter stereoscopic range finder set up nearby. The super elevation and fuze are taken from three-dimensional cams, positioned by future angular height and future slant range. Lateral, vertical, and range rates are measured by tachometers and are manually matched. Deflections are computed by multiplying present angular velocity by present time of flight.

(3) The instrument and stand are carried in a two-wheeled trailer. The stand is collapsible for traveling. Three leveling screws on the head of the stand support the director. The director has four porter bars which form an integral part of the instrument and telescope within each other when the instrument is emplaced. The director weighs about 450 pounds.

(4) The only electric current required is for the telephone system. The 10-pole receptacle in the center of the bottom provides constant contact throughout complete traversing of the director because of the slip rings in the bottom of the director. The receptacle has colored dots: black, white, yellow, brown, brown, respectively for adjacent pairs of terminals. A key prevents incorrect positioning of the mating plug. The receptacles on the gun mount use colors in accordance with this color system. The two receptacles also at the bottom, but on either side of the center, move with the director and permit connection to the telephone sets for the director operators.

b. Operation.

(1) Limits of operation are:

Range of target.................................... 12,000 meters (39,360 ft)

Height of target.................................... 10,000 meters (32,800 ft)

Fuze time .... 37 seconds

Lateral deflection .... $\pm$ 600 mils (33.75 deg)

Vertical deflection .... $\pm$ $^{300}/_{16}$-degrees (18.75 deg)

Rate of change of range.... $\pm$ 150 meters per second (492 ft per sec)

GERMAN 88-MM ANTIAIRCRAFT GUN MATERIEL

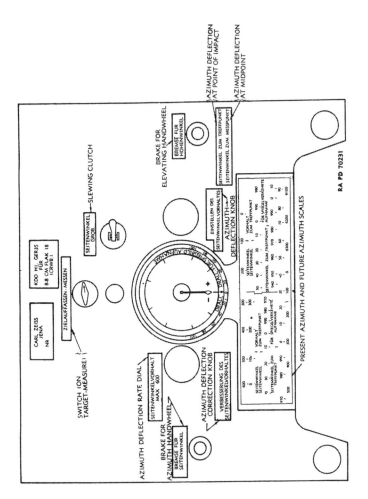

Figure 96 — Auxiliary Director 35 (Kdo. Hi. Gr. 35) — Rear Panel

134

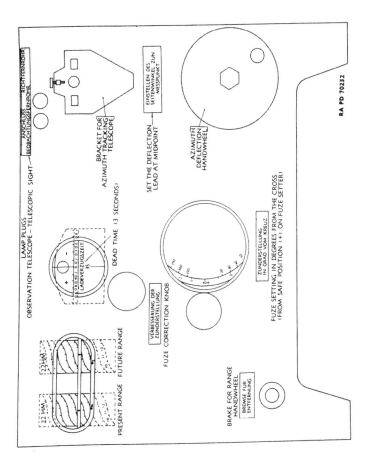

RA PD 70232

Figure 97 — Auxiliary Director 35 (Kdo. Hi. Gr. 35) — Left Side Panel

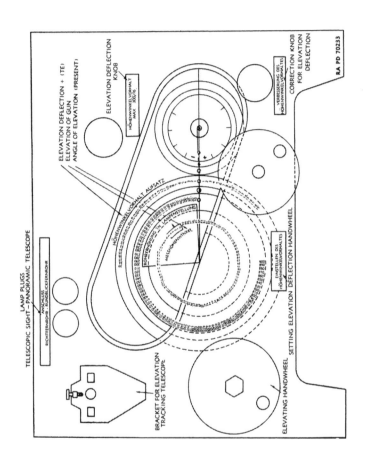

Figure 98 — Auxiliary Director 35 (Kdo. Hi. Gr. 35) — Right Side Panel

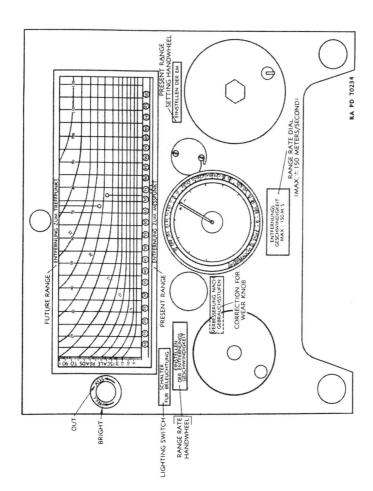

Figure 99 — Auxiliary Director 35 (Kdo. Hi. Gr. 35) — Front Panel

137

GERMAN 88-MM ANTIAIRCRAFT GUN MATERIEL

(2) It is believed that nine operators are required:

(a) Azimuth tracker.

(b) Elevation tracker.

(c) Range operator.

(d) Azimuth rate setter.

(e) Elevation rate setter.

(f) Range rate setter.

(g) Future azimuth reader.

(h) Future elevation reader.

(i) Fuze reader.

(3) The azimuth and elevation trackers keep their telescopes on the target by operating their respective handwheels. The range operator sets in range values obtained from the 4-meter stereoscopic range finder nearby.

(4) The azimuth rate setter keeps the azimuth rate indicator alined with its index (at 6 o'clock relative to the rate dial for zero correction) by turning the azimuth deflection knob. He also applies correction by turning the azimuth rate knob.

(5) The elevation rate setter alines the elevation rate pointer with its index (positioned under the white line for zero corrections) and applies corrections by turning the elevation rate correction knob.

(6) The range rate setter keeps the range rate pointer alined with the index (at 6 o'clock for zero correction) and applies corrections by turning the range rate correction knob.

(7) The future azimuth reader observes the future azimuth scale and telephones the values indicated to the guns. The future elevation reader observes the future elevation scale on the right side panel and telephones the values indicated to the guns. The fuze reader observes the fuze dial on the left side panel and telephones the values to the guns.

(8) Each input handwheel has a flywheel mounted within the case which tends to give a steady rate of change. Brakes operated by push buttons permit quick stopping.

(9) Clutches are provided for driving in azimuth or elevation.

79. AZIMUTH AND ELEVATION INDICATORS.

a. Description.

(1) The azimuth and elevation indicators on the gun carriage are identical (fig. 100). Each indicator consists of a cylindrical case containing three concentric rings of 10-volt electric lamp sockets. Each socket has an individual positive electrical connection. All sockets in each indicator have a common negative connection.

(2) Three pointers are pivoted at the center of each indicator, one for each circle of lights. At the end of each pointer is a translucent index. The two inner indexes are each wide enough to cover one light. The outer pointer can span two lights. The pointers are

FIRE CONTROL EQUIPMENT

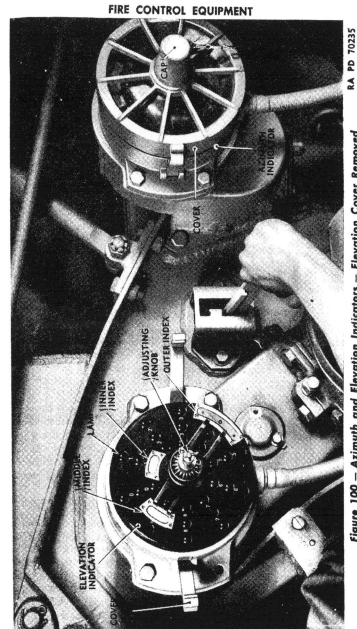

RA PD 70235

Figure 100 — Azimuth and Elevation Indicators — Elevation Cover Removed

139

GERMAN 88-MM ANTIAIRCRAFT GUN MATERIEL

RA PD 70236

Figure 101 — Fuze Setter on Carriage

geared together at a ratio of 1:10:100; the shortest pointer moving 1 turn to 100 of the target pointer. The pointers are mechanically coupled to the azimuth and elevation drives of the gun carriage. A knob at the center of each indicator is used for synchronizing the indicator arms with the gun prior to operation.

(3) The case is bolted to the top carriage and has a cover which consists of a 10-armed spider which supports a translucent plastic sheet cupped within the spider. A dowel on the case fits a hole in the rim of the spider, permitting assembly in only one position. This assures maximum visibility of the lights. Two spring clips clamp the cover in place. At the center of the spider is a chained cap which protects the adjusting knob of the indicator.

b. Operation.

(1) Sight the range finder of the stereoscopic director on a distant aiming point (a distant terrestrial object or a celestial body) and bore sight the guns of the battery on this target. If the lighted lamps of the azimuth and elevation indicators of each gun cover the lighted lamps the gun is oriented with the director. If adjustment is necessary, remove the metal cap and engage the knob with the cross piece in the shaft, turn until the dial is blacked out, and release the knob.

(2) The gun is thereafter operated as the lights flash on around the circle by turning the elevating and traversing handwheels to keep the lamps blacked out.

FIRE CONTROL EQUIPMENT

CABLE AND PLUG

FUZE CUP

LOCKING PIN
RELEASE LEVER

RA PD 70237

CLUTCH

TIME SETTING
HANDWHEEL

FUZE SCALE

FUZE SETTING CRANK (INERTIA)

Figure 102 — Fuze Setter — Close-up, Showing Cover Removed and One Fuze Cap Disassembled

GERMAN 88-MM ANTIAIRCRAFT GUN MATERIEL

80. FUZE SETTER.

a. Description.

(1) The fuze setter is mounted on the left side of the top carriage (fig. 101). It is manually operated and is capable of cutting two fuzes at a time. It may be used with either the stereoscopic director or the auxiliary director.

(2) For use with data electrically transmitted from the stereoscopic director, the fuze setter has a system of lights similar to that employed in the azimuth and elevation indicators. In the top of the fuze setter is a plate containing three concentric rings of 10-volt electric lamp sockets. Each socket has an individual positive electrical connection. All the sockets have a common negative connection. Three pointers are pivoted at the center of the circles, one for each circle of lights. At the end of each pointer is a translucent index. The two inner indexes are each wide enough to cover one light. The outer pointer can span two lights. The pointers are geared together at a ratio of 1:10:100, the shortest pointer moving 1 turn to 100 turns of the longest pointer. The pointers are geared to the fuze dial and to the setting ring of the fuze setter.

(3) For use with data telephoned from the auxiliary director, the fuze setter has a scale graduated from 15 to 350 degrees (fig. 102). The safe position is marked with a cross. (American fuze setter dials are marked in fuze seconds. A conversion scale is necessary for converting from American fuze seconds, with corrector values set in to the degree markings on this fuze setter.)

(4) The setting crank at the front of the fuze setter turns the pointers and the fuze dial. The crank at the side of the fuze setter turns the inertia flywheel which stores up mechanical energy for cutting the fuzes. The release lever releases the round after the fuze is cut. The cable receptacle from the fuze setter extends to the terminal box at the front of the top carriage.

b. Operation.

(1) Turn the setting handwheel to black out the lights (in operation with the stereoscopic director) or to aline the fuze scale in accordance with the values announced by telephone (in operation with the auxiliary director). Turn the power crank, thus storing up energy in the flywheel, and keep the crank turning at a uniform rate.

(2) Thrust the round sharply into one of the cups of the fuze setter, thereby engaging a toothed clutch which rotates the adjusting pin. This makes two complete turns before being automatically disengaged. The round is held in position by a key which rides in a circular groove at the bottom of the fuze. This key is tripped by a lever at the top of the setter and the round is released. Two fuzes may be cut at one time.

FIRE CONTROL EQUIPMENT

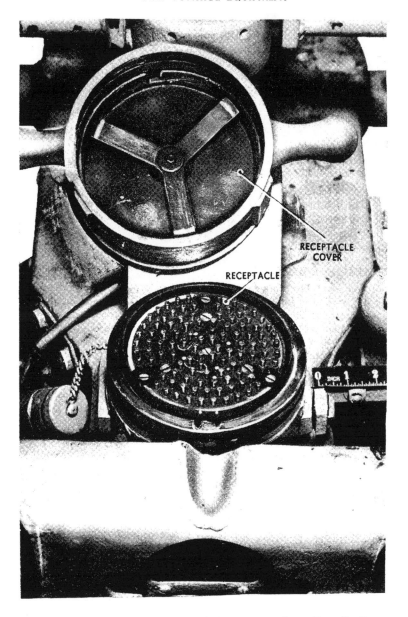

Figure 103 — Data Transmission Receptacle on Rear Trail

143

GERMAN 88-MM ANTIAIRCRAFT GUN MATERIEL

81. ON-CARRIAGE WIRING.

a. A receptacle containing 104 pins is at the end of the rear trail and is intended for connection to either of the directors (fig. 103). Each pin is numbered and groups of pins are identified by colored dot inserts. Conductors are brought from the end of the trail to a receptacle box at the front of the top carriage. The arrangement of wiring permits traversing the gun a maximum of two turns in either direction. Stops limit further travel and a dial (marked "MUNDUNG," meaning muzzle direction) just over the traversing handwheel is graduated to indicate the number of turns made.

b. Cables and plugs from the azimuth and elevation indicators, the fuze setter and the receptacle, adjacent to each, plug into the receptacle box.

c. The bell just above the fuze setter may be sounded to indicate that the director is on target or may serve as a time interval bell.

d. Conductors are carried in loose flexible coverings throughout the gun carriage and are not armored.

82. AIMING CIRCLE.

a. General. The aiming circle (figs. 104 to 107) is used for measuring angle of site, for declinating and determining azimuth angles, and for spotting. The instrument, removed from the tripod, may be used on a plane table for topographic survey. The aiming circle consists of a periscope, a telescope having 4- or 5-power magnification, an angle of site mount, an azimuth mount, and a tripod. The tripod is the same as that of the battery commander's telescope. Carrying cases for the instrument and the tripod are provided. A trench mount is furnished which can be embedded in the ground or in wood for use in place of the tripod. A lamp bracket and portable battery supplies light for the telescope reticle. Graduations are in mils.

b. Description of Components.

(1) The periscope raises the line of sight, but has no magnifying power. It is attached to the aiming circle by a dovetailed slide, but is not locked in place. The aiming circle may be used without the periscope.

(2) The telescope has an adjustable focusing eyepiece. Horizontal and vertical cross lines and a deflection scale are on the reticle of the telescope. On top of the telescope body is a level used with the angle of site mechanism. On the left of the telescope is a circular level. A sun shade is provided for use when the periscope is not attached.

(3) The angle of site mount supports the telescope and includes a graduated elevation scale and micrometer, a magnetic needle, a circular level, and clamping devices.

(a) The elevation scale, graduated from 0 to 1,400 mils, and the micrometer 0 to 100 mils. The normal setting is 300 mils.

FIRE CONTROL EQUIPMENT

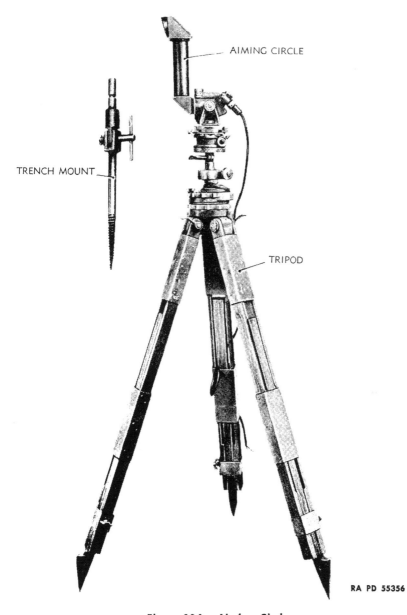

AIMING CIRCLE

TRENCH MOUNT

TRIPOD

RA PD 55356

Figure 104 — Aiming Circle

GERMAN 88-MM ANTIAIRCRAFT GUN MATERIEL

TRANSLATION

LOS - LOOSE

FEST - LOCKED

OSTW. - EASTWARD

WESTI. - WESTWARD

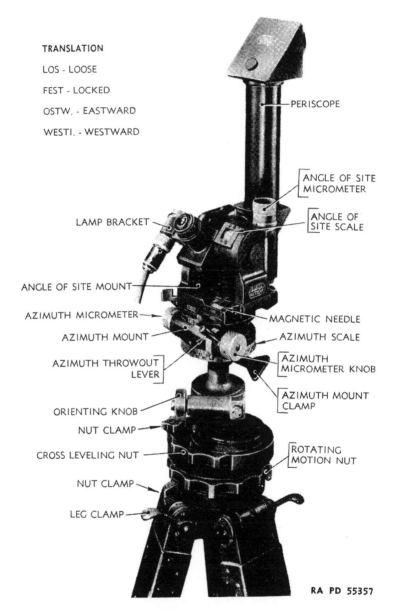

PERISCOPE

ANGLE OF SITE MICROMETER

ANGLE OF SITE SCALE

LAMP BRACKET

ANGLE OF SITE MOUNT

AZIMUTH MICROMETER

MAGNETIC NEEDLE

AZIMUTH MOUNT

AZIMUTH SCALE

AZIMUTH THROWOUT LEVER

AZIMUTH MICROMETER KNOB

AZIMUTH MOUNT CLAMP

ORIENTING KNOB

NUT CLAMP

CROSS LEVELING NUT

ROTATING MOTION NUT

NUT CLAMP

LEG CLAMP

RA PD 55357

Figure 105 — Aiming Circle — Close-up

CARRYING CASE

TRANSLATION

⌈VOR EINSETZEN DES RKR.
⌊MAGNETNA DEL FESTLAGEN!

BEFORE PACKING THE
AIMING CIRCLE CLAMP
THE MAGNETIC NEEDLE

⌈AUSBLICK - CAUTION
⌈INHALTSVERZEICHNIS
⌊LIST OF CONTENTS

⌈KASTEN RKR 31
⌊CHEST - AIMING CIRCLE 31

⌈1 - RICHTKREIS -
⌊1 AIMING CIRCLE
⌈1 - DECKUNGSPIEGEL -
⌊1 PERISCOPE
⌈1 - ANSTECKLAMPE -
⌊1 PLUG IN LAMP
⌈1 - BEHALTER FUR
STROMQUELLE -
⌊1 BATTERY BOX
⌈1 - KLARINOLTUCH IN
TACHE - 1 IMPREGNATED
⌊CLOTH
⌈4 GLUHLAMPEN - 4 LAMPS
⌊3.5 V, 0.2 A.
⌈1 PUTZTUCH -
⌊1 CLEANING CLOTH
⌈1 STAUBPINSEL -
⌊1 DUST BRUSH

AZIMUTH MOUNT

PERISCOPE

ANGLE OF SITE MOUNT

RA PD 55358

Figure 106 — Aiming Circle Components

GERMAN 88-MM ANTIAIRCRAFT GUN MATERIEL

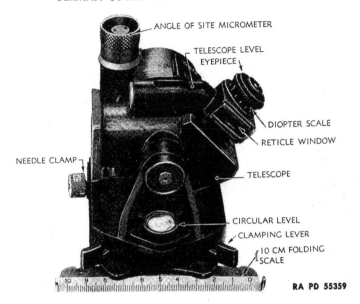

Figure 107 — Aiming Circle — Angle of Site Mount

(b) The magnetic needle has a visible range of 10 degrees on either side of the magnetic north line. A knob below the window marked "N" locks the needle when it is not in use. A window at the "S" end permits observation from the rear of the instrument.

(c) The circular level is used with the compound head of the tripod for leveling the instrument.

(d) Clamping levers lock the angle of site mount on the azimuth mount.

(e) The folded 10-centimeter ruler (fig. 107), is for use when the telescope and angle of site mount, disengaged from the traversing mechanism, is in use on a plane table in topographic survey.

(4) The azimuth mount has a tapered stud which supports the angle of site mount, an azimuth scale, graduated from 0 to 6,400 mils in 100-mil intervals, and a micrometer, graduated from 0 to 100 mils in 1-mil intervals. A throw-out lever permits rapid traversing of the instrument. The azimuth mount is clamped to the spindle of the tripod.

(5) The tripod is used for both the aiming circle and battery commander's telescope. The tripod includes a spindle, a worm and worm wheel mechanism, a ball and socket joint, and individually clamped legs. The spindle supports the instrument and is attached to the worm and worm wheel mechanism which is used for orientation.

FIRE CONTROL EQUIPMENT

The ball and socket joint includes the ball at the end of the spindle, two clamping nuts, one of which permits cross leveling and the other, circular oscillation. The tripod legs have clamping levers at the head for locking each leg to the head. At the foot of each leg is a steel point and foot rest which facilitate embedding in the ground.

(6) The carrying case is provided for the instrument. The table of contents pasted in the cover includes one lamp bracket and four lamps, a dry cell battery holder, a dust brush, a cleaning cloth, and an impregnated cloth to be used in decontaminating parts of the instrument which may have been subjected to gas attack.

c. Operation.

(1) To set up the instrument, clamp the tripod legs at the desired length and embed them firmly in the ground. Level the instrument using the circular level and the ball and socket joint. Tighten the clamping nuts. Focus the telescope as required, using the sleeve on the eyepiece to set in the correction necessary for the observer's eye.

(2) To orient the instrument, a datum point of known azimuth or a magnetic bearing may be used.

(a) To orient on a datum point of known azimuth, set the main azimuth scale (100-mil intervals) and micrometer (1-mil intervals) to the azimuth of the datum point and turn the orienting knob until the datum point appears on the vertical cross line of the reticle. The instrument may also be relocated on the tripod spindle using the orienting clamping screw for large angular changes. The telescope may be elevated or depressed as required to center the point in the field of view.

(b) To orient on magnetic north, set the main azimuth scale and micrometer to indicate zero. Press the plunger releasing the magnetic needle and turn one of the orienting knobs until the north-seeking end of the magnetic needle appears approximately opposite the "N" index at the front of the instrument; then refine the setting so that the south-seeking end of the needle is centered in the reticle. The instrument may also be relocated on the tripod spindle using the orienting clamping screw for large angular changes. The aiming circle will then indicate magnetic azimuths.

(c) To orient on grid north, proceed as for magnetic north but set the azimuth to the magnetic declination of the locality (subtracting west declinations from 6,400 mils) instead of to zero. The instrument will then indicate grid azimuths.

(d) When orientation by magnetic bearings has been completed, turn the knob to clamp the magnetic needle.

(3) To read azimuth, bring the object on the vertical cross line of the reticle using the azimuth knob; the throw-out lever may be depressed for making large azimuth changes rapidly. Azimuths from

GERMAN 88-MM ANTIAIRCRAFT GUN MATERIEL

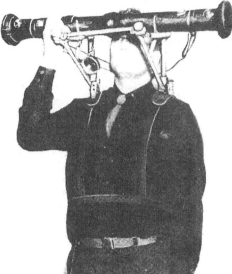

RA PD 55360

Figure 108 — Range Finder 34 with Harness

0 to 6,400 mils are read directly on the azimuth scale; the scale is graduated at 100-mil intervals and the micrometer is graduated at 1-mil intervals.

(4) To read angle of site, first make sure that with the telescope level bubble and the circle level bubble centered the angle of site scale and micrometer read normal. Then center the object in azimuth. Raise or depress the angle of site micrometer to center the object on the reticle cross lines and read the angle of site on the scale and micrometer.

(5) To prepare the instrument for traveling, place it in the carrying case provided.

d. **Tests and Adjustments.**

(1) The azimuth micrometers should read "0" when the azimuth scale indicates zero. The screw in the end of the azimuth micrometer may be temporarily loosened for this adjustment.

(2) The line of sight as determined by the center of the reticle should be horizontal when the bubble in the telescope level is centered. This may be verified by sighting on a distant point at the same level as the telescope, the error, if any, being read on the reticle. No corrective adjustment by the using arms is permitted.

FIRE CONTROL EQUIPMENT

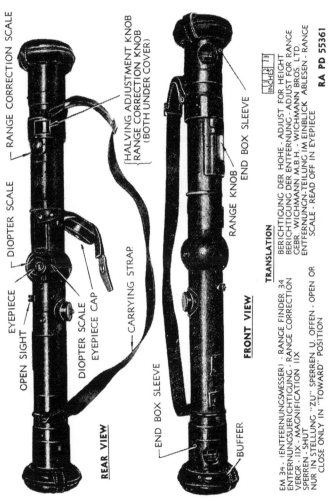

RANGE CORRECTION SCALE

HALVING ADJUSTMENT KNOB
RANGE CORRECTION KNOB
(BOTH UNDER COVER)

DIOPTER SCALE

END BOX SLEEVE

EYEPIECE

DIOPTER SCALE

EYEPIECE CAP

OPEN SIGHT

RANGE KNOB

CARRYING STRAP

END BOX SLEEVE

BUFFER

INCHES

RA PD 55361

REAR VIEW

FRONT VIEW

TRANSLATION

EM 34 - (ENTFERNUNGSMESSER) - RANGE FINDER 34   BERICHTIGUNG DER HÖHE - ADJUST FOR HEIGHT
ENTFERNUNGSUERICHTIGUNG - RANGE CORRECTION   BERICHTIGUNG DER ENTFERNUNG - ADJUST FOR RANGE
VERGR. - IIX - MAGNIFICATION IIX   GEBR. WICHMANN M.B.H. - WICHMANN BROS. LTD.
SPERREN - SHUT   ENTFERNUNGN-TEILUNG IM EINBLICK ABLESEN - RANGE
NUR IN STELLUNG "ZU" SPERREN U. OFFEN - OPEN OR   SCALE - READ OFF IN EYEPIECE
    CLOSE ONLY IN "TOWARD" POSITION

Figure 109 – Range Finder 34 – Assembled Views

151

GERMAN 88-MM ANTIAIRCRAFT GUN MATERIEL

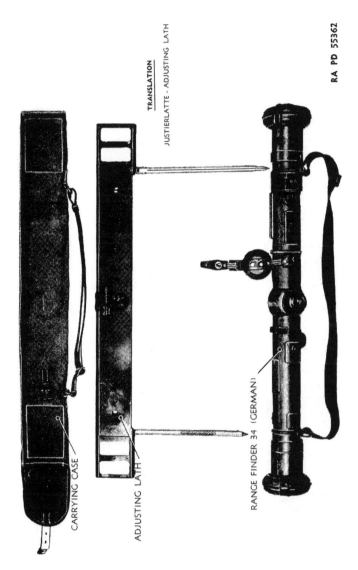

TRANSLATION
JUSTIERLATTE - ADJUSTING LATH

RA PD 55362

CARRYING CASE

ADJUSTING LATH

RANGE FINDER 34 (GERMAN)

Figure 110 — Range Finder 34 with Adjusting Lath

FIRE CONTROL EQUIPMENT

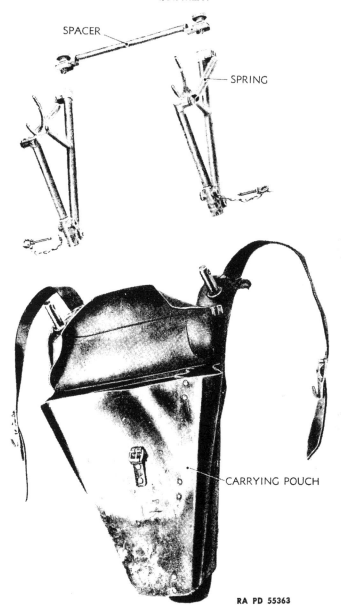

RA PD 55363

Figure 111 — Range Finder 34 — Harness

GERMAN 88-MM ANTIAIRCRAFT GUN MATERIEL

(3) To check the accuracy of the declinator, it is necessary to set up the instrument in a position not subject to local magnetic attraction and sight on one or (preferably) more points of known azimuth. The average error should be noted and the necessary correction recorded. No adjustment by the using arms is permitted.

e. **Care and Preservation.** Refer to paragraph 87 for general instructions pertaining to the care and preservation of instruments.

83. RANGE FINDER MODEL 34.

a. This instrument (figs. 108 to 111) is used primarily for measuring distances by triangulation. Range values are read in the field of view.

b. **Description.** The instrument includes an internal 70-centimeter base line, all power optical system with two objectives, a common eyepiece of the coincidence type, and a scale on which the distance is indicated. It is furnished complete with a carrying strap, an eyepiece cap and strap, a shoulder harness with carrying pouch, and an adjusting lath with carrying case.

c. **Operation.**

(1) To set up the instrument, adjust the harness on the observer. See figure 108. The carrying pouch should hang on the back; the spring mounted holders for the range finder should extend in front of the observer. Carefully place the range finder in the holders and in line with the eye of the observer.

(2) Focus the eyepiece by rotating the diopter scale to produce a sharp image; if the operator knows the value for his own eye, the setting may be made directly on the scale.

(3) To measure the range of a target, aline the instrument on the target, using the open sight. Select a clearly defined part, perpendicular if possible to the halving line. Center the target in the field of view. Turn the range knob until the images of the target appear in coincidence. Read the range value centered in the field of view.

(4) To prepare the instrument for traveling, remove the instrument from the harness, close the end box covers, and cover the eyepiece. Disassemble the harness and put it in the carrying pouch.

d. **Tests and Adjustments.** Information on this is not available at this time but will be published when available.

e. **Care and Preservation.**

(1) Refer to paragraph 87 on general instructions pertaining to the care and preservation of instruments.

(2) Keep the end box sleeves closed and eyepiece covered when the instrument is not in use.

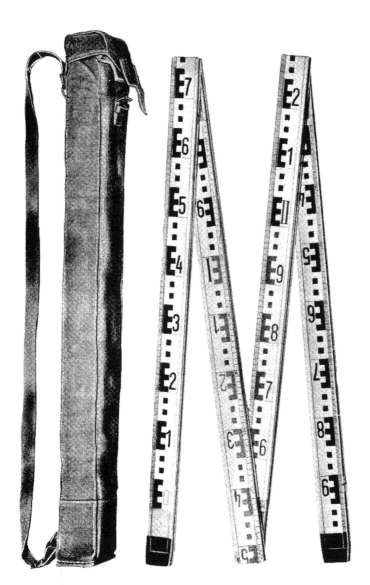

RA PD 55364

Figure 112 — Surveying Rod and Carrying Case

GERMAN 88-MM ANTIAIRCRAFT GUN MATERIEL

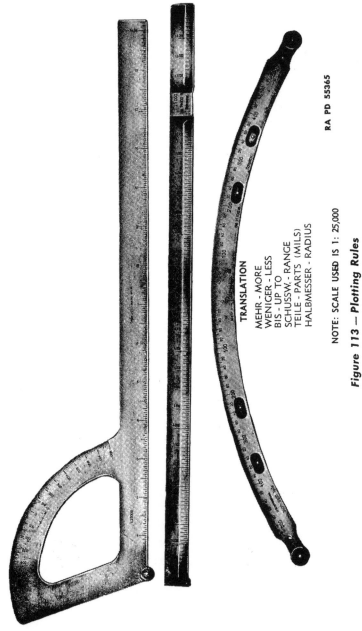

RA PD 55365

TRANSLATION

MEHR - MORE
WENIGER - LESS
BIS - UP TO
SCHUSSW.- RANGE
TEILE - PARTS (MILS)
HALBMESSER - RADIUS

NOTE: SCALE USED IS 1: 25,000

Figure 113 — Plotting Rules

FIRE CONTROL EQUIPMENT

84. SURVEYING ROD.

a. A 3-meter surveying rod (fig. 112), graduated in 1-centimeter divisions, is provided for orientation of the battery. The rod is hinged to reduce its length to about three-quarters of a meter for storage and travel. The folded rod is carried in a canvas case fitted with a sling strap.

85. PLOTTING RULES.

a. Three rules are furnished as plotting board accessories.

b. One steel rule (fig. 113) bears a linear scale, graduated from 0 to 14 kilometers in 5-meter intervals, and a quadrant, graduated from minus 800 mils to plus 800 mils in 50-mil intervals. At the zero end is a center for pivoting. This rule may be used for plotting azimuth and range values on a topographic map with a scale of 1:25000.

c. The steel protractor is graduated from minus 500 mils to plus 500 mils in 2-mil intervals. The radius of curvature is 480 milimeters. Two points permit pinning the protractor to a board.

d. The other steel rule bears a linear scale graduated from 0 to 14,600 meters in 50-meter intervals and a center at the zero end. At about 11,800 meters is a raised section. When this rule is set up with the protractor on a deflection chart, the raised section clears the protractor. The rule and protractor are used for plotting deflections on a chart with a scale of 1:25000.

86. BATTERY COMMANDER'S TELESCOPE.

a. Description.

(1) The battery commander's telescope is a 10-power binocular instrument used for observation and for measuring azimuths and angles of site. The instrument consists of a telescope and an azimuth mount, tripod, carrying case, and accessories. The tripod includes an orienting mount. A trench mount is furnished, which can be embedded in the ground or in wood, for use in place of the tripod.

(2) The telescope arms may be positioned vertically (fig. 114) or they may be swung horizontally (fig. 116) to increase the steroscopic effect. The reticle, which remains erect in any position of the telescope arms, is illuminated by the removable lamp on the slide near the reticle.

b. Operation.

(1) To set the instrument, clamp the tripod legs at the desired length, embed them firmly in the ground, and tighten the leg clamping levers. Using the spring plunger, clamp the telescope on the vertical spindle extending from the orienting mount. (The tripod has a mount which permits cross leveling and orienting.) Level the mount by cen-

GERMAN 88-MM ANTIAIRCRAFT GUN MATERIEL

BC. TELESCOPE

TELESCOPE
CLAMPING
KNOB

INTERPUPILLARY CLAMP KNOB

TRENCH
MOUNT

TRANSLATION

1 S.F. - B.C. TELESCOPE, SCISSOR TYPE
1 MEBKREIS - MOUNT, ON TRIPOD
1 EINGELENKBAUMS CHRAUBE - AZIMUTH
  MOUNT
2 REGENSCHUTZOHRE - 2 RAIN PROTECTOR
  TUBES
1 BEHALTER FUR STROMQUELLE - CONTAINER
  FOR BATTERY
1 HANDLAMPE - HAND LAMP
1 ANSTECKLAMPE - PLUG IN LAMP
1 VERTEILER, ZWEIFACH - DOUBLE
  DISTRIBUTOR
4 GLUHBIRNEN 3.5V (Z. VORRAT) - 4 BULBS
  3.5 V.
1 AUGENMUSCHEL (Z. VORRAT) - EYE GUARD
2 BLANDGLASER, HELL - 2 FILTERS, LIGHT
2 BLENDGLASER, DUNKEL - 2 FILTERS, DARK
1 PUTZUCH - CLEANING CLOTH
1 STAUBPINSEL - DUST BRUSH
1 KLARINOLTUCH IN TASCHE - IMPREGNATED
  CLOTH IN PACKET

B.C. TELESCOPE
CARRYING CASE

TRIPOD
CARRYING
CASE

TRIPOD

RA PD 55368

Figure 114 — Battery Commander's Telescope with Cases

158

FIRE CONTROL EQUIPMENT

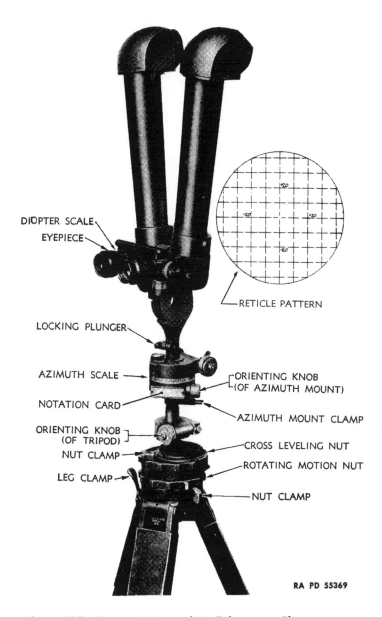

DIOPTER SCALE
EYEPIECE

RETICLE PATTERN

LOCKING PLUNGER

AZIMUTH SCALE

ORIENTING KNOB
(OF AZIMUTH MOUNT)

NOTATION CARD

AZIMUTH MOUNT CLAMP

ORIENTING KNOB
(OF TRIPOD)

NUT CLAMP

CROSS LEVELING NUT

LEG CLAMP

ROTATING MOTION NUT

NUT CLAMP

RA PD 55369

Figure 115 — Battery Commander's Telescope — Close-up

GERMAN 88-MM ANTIAIRCRAFT GUN MATERIEL

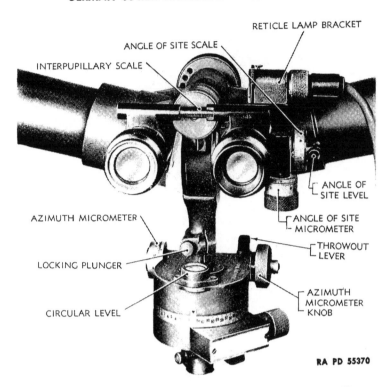

*Figure 116 — Battery Commander's Telescope — Interpupillary
Scale and Angle of Site Mechanism*

tering the bubble in the circular level. When the bubble is centered,
clamp the ball and socket joint on the lower mount.

(2) To prepare the telescope, release the telescope clamping knob
(fig. 114) and turn the telescope arms to the vertical or horizontal
position, as required. Set the proper interpupillary distance on the
interpupillary scale (fig. 115), graduated from 55 to 75 milimeters,
and tighten the interpupillary clamp knob. If the interpupillary distance for the observer is not known, it may be found by observing the
sky and moving the eyepiece apart or together until the field of view
changes from two circles or two overlapping circles to one sharply
defined circle. The interpupillary wing knob is then clamped. Focus
each eyepiece independently by covering one of them and looking
through the telescope with both eyes open at an object several hun-
dred yards away; turn the diopter scale until the object observed
appears sharply defined. The diopter scale on each eyepiece permits

FIRE CONTROL EQUIPMENT

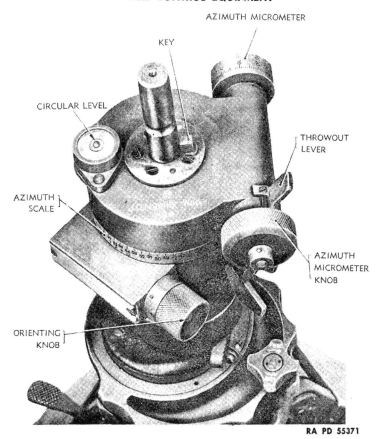

AZIMUTH MICROMETER

KEY

CIRCULAR LEVEL

THROWOUT
LEVER

AZIMUTH
SCALE

AZIMUTH
MICROMETER
KNOB

ORIENTING
KNOB

RA PD 55371

Figure 117 — Battery Commander's Telescope — Azimuth Mount

immediate adjustment for each eye if the observer knows his own eye
corrections. If required, place the light or dark filters over the eyepieces and the metal sunshades over the objective lenses. Tubular
sections, about 8 inches long, can be attached to the sun shades for
protection against rain.

(3)   To orient the instrument, select a datum point of known
azimuth and set this value on the azimuth scale (100-mil steps) and
micrometer (1-mil steps). The throw-out lever may be used for making
large changes in azimuth rapidly. Turn the telescope with the orienting
knob until the datum point appears at the center of the reticle of the
right-hand telescope. The orienting clamping knob may be temporarily
released for making large angular changes rapidly. Thereafter, use

only the azimuth knob, or, for large changes, the azimuth throw-out lever, and the correct azimuth of the point observed will be indicated. For azimuths in the 3,200- to 6,400-mil region additional numbers (0 to 3,200 mils) are provided, corresponding to the azimuth scale on the panoramic telescope.

(4)   Direct the telescope on the object and rotate the elevating knob until the object appears at the center of the reticle. Center the angle of site level bubble by adjusting the angle of site knob. The angle of site is then read on the angle of site scale. It is not known whether the graduations on the angle of site scale are in mils or in $\frac{1}{16}$ degrees. An indication of 300 corresponds to a horizontal line of sight.

(5)   The reticle (fig. 115), located in the right eyepiece, is a grid, the horizontal and vertical axis of which are graduated in 100-mil intervals.

(6)   A throw-out mechanism is provided for rapidly traversing the telescope. A circular level is provided for leveling the head. The traversing head is graduated from 0 to 64 in 100-mil divisions with a micrometer adjustment from 0 to 100 in 1-mil divisions.

(7)   To prepare the instrument for traveling, remove the sun shades and filters, if used. Loosen the telescope clamping knob and place the telescope arms in a vertical position. Disengage the telescope from the mount and place the instrument in the wooden carrying case.

c.   **Tests and Adjustments.**

(1)   The azimuth micrometer and azimuth scale should read zero simultaneously. The screw in the end of the micrometer may be temporarily loosened to permit adjustment.

(2)   The angle of site mechanism may be checked by observing a datum point of known angle of site. Small errors may be corrected by temporarily loosening the screw in the end of the knob and slipping the micrometer and knob to the correct position. Should the angle of site scale and micrometer then fail to indicate "300" and "0" respectively, the instrument should be turned in for adjustment by authorized ordnance personnel.

d.   **Care and Preservation.**

(1)   Refer to paragraph 87 for general instructions pertaining to the care and preservation of instruments.

(2)   Always release the telescope clamping knob before rotating the telescopes in a vertical plane. Failure to do this often results in damaging the instrument and causes double vision.

FIRE CONTROL EQUIPMENT

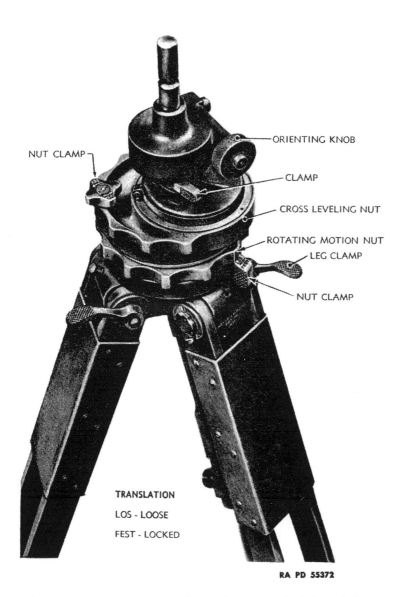

ORIENTING KNOB

NUT CLAMP

CLAMP

CROSS LEVELING NUT

ROTATING MOTION NUT

LEG CLAMP

NUT CLAMP

TRANSLATION

LOS - LOOSE

FEST - LOCKED

RA PD 55372

*Figure 118 — Battery Commander's Telescope and Aiming Circle —
Tripod Head*

GERMAN 88-MM ANTIAIRCRAFT GUN MATERIEL

87. CARE AND PRESERVATION.

a. General.

(1) The instructions given hereunder supplement instructions pertaining to individual instruments included in preceding paragraphs.

(2) Fire control and sighting instruments are, in general, rugged and suited for the purpose for which they have been designed. They will not, however, stand rough handling or abuse. Inaccuracy or malfunctioning may result from such mistreatment.

(3) Unnecessary turning of screws or other parts not incident to the use of the instrument is expressly forbidden.

(4) Keep the instruments as dry as possible. Do not put an instrument in its carrying case when wet.

(5) When not in use, keep the instruments in the carrying cases provided, or in the condition indicated for traveling.

(6) The maintenance duties described are those for which tools and parts have been provided the using personnel. Other replacements and repairs are the responsibility of maintenance personnel, but may be performed by the using arm personnel, when circumstances permit, within the discretion of the commander concerned.

(7) No painting of fire control or sighting equipment by the using arms is permitted.

(8) Many worm drives have throw-out mechanisms to permit rapid motion through large angles. When using these mechanisms, it is essential that the throw-out lever be fully depressed to prevent injury to the worm and gear teeth.

(9) When using a tripod with adjustable legs, be certain that the legs are clamped tightly to prevent possibility of collapse.

(10) When setting up tripods on sloping terrain, place two legs on the downhill side to provide maximum stability.

(11) Dry cell batteries should not be kept in the battery boxes when the instrument is not in use. Dry cell batteries when weak deteriorate rapidly and will cause corrosion and other damage to containers.

(12) Data transmission cables should be protected against crushing by vehicles.

b. Optical Parts.

(1) To obtain satisfactory vision, it is necessary that the exposed surfaces of the lenses and other parts be kept clean and dry. Corrosion

FIRE CONTROL EQUIPMENT

and etching of the surface of the glass can be prevented or greatly retarded by keeping the glass clean and dry.

(2) Under no condition will polishing liquids, pastes, or abrasives be used for polishing lenses and windows.

(3) For wiping optical parts, use only lens paper specially intended for cleaning optical glass. Use of cleaning cloths in the field is not permitted. To remove dust, brush the glass lightly with a clean camel's-hair brush and rap the brush against a hard body in order to knock out the small particles of dust that cling to the hairs. Repeat this operation until all dust is removed. With some instruments an additional brush with coarse bristles is provided for cleaning mechanical parts; it is essential that each brush be used only for the purpose intended.

(4) Exercise particular care to keep optical parts free from oil and grease. Do not wipe the lenses or windows with the fingers. To remove oil or grease from optical surfaces, apply ALCOHOL, ethyl, grade 1, with a clean camel's-hair brush and rub gently with clean lens paper. If alcohol is not available, breathe heavily on the glass and wipe off with clean lens paper; repeat this operation several times until clean.

(5) Moisture due to condensation may collect on the optical parts of the instrument when the temperature of the parts is lower than that of the surrounding air. This moisture, if not excessive, can be removed by placing the instrument in a warm place. Heat from strongly concentrated sources should not be applied directly, as it may cause unequal expansion of parts, thereby resulting in breakage of optical parts or inaccuracies in observation.

c. **Lubricants.**

(1) Where lubrication with oil is indicated, use OIL, lubricating, for aircraft instruments and machine guns.

(2) Where lubrication with grease is indicated, use GREASE, lubricating, special.

(3) Exposed moving points should be oiled occasionally. Interior parts are not to be lubricated by the using arms. Wipe off any excess lubricant that seeps from the mechanisms to prevent accumulation of dust and grit.

(4) The tripod pivots should be carefully oiled at frequent intervals.

(5) Do not oil optical parts.

GERMAN 88-MM ANTIAIRCRAFT GUN MATERIEL

CHAPTER 5

FIRING TABLES

*Kz—Kopfzünder = Nose fuze.

FIRING TABLES

88. FIRING TABLE FOR THE 8.8 cm FLAK 18 AND 36 WITH 8.8 cm H.E. SHELL L/4.5 (Kz)* WITH TIME FUZE S/30 OR P.D. FUZE 23/28 (TABLE I).

M.V. 820 m/s or 2690 ft/sec.   Weight of shell 9.00 kg or 19.8 lb   Weight of air at ground level = 1.22 kg/m³ or 533 grains/cu ft

Range		Quadrant Elevation	Drift	For S/30 Fuze Only Fuze Setting at 0 Meters Remaining Flight	Time of Flight	Maximum Ordinate		1/16 deg Changes the Point of Impact by	Means (50 percent) Dispersion			Angle of Impact	Velocity
						Distance	Height		Range	Width	Height		
yd	m	deg	mils	deg from cross	sec	m	m	m	m	m	m	deg	m/s
1	1	2*	3	4	5	6	7	8	9	10	11	12*	13
109	100	0 1	0	—	0.13	50	0	147	—	—	—	0 1	810
219	200	0 1	0	—	0.26	100	0	145	68	0.1	0.2	0 1	800
328	300	0 2	0	—	0.39	150	1	143	68	0.1	0.3	0 2	789
437	400	0 3	0	21	0.52	201	1	140	69	0.1	0.3	0 2	779
547	500	0 4	0	22	0.65	252	2	138	69	0.2	0.4	0 3	769
656	600	0 4	0	24	0.78	303	2	136	69	0.2	0.4	0 4	759
766	700	0 5	0	25	0.91	354	3	134	69	0.2	0.5	0 5	749
875	800	0 6	0	27	1.05	406	3	131	69	0.3	0.6	0 6	739
984	900	0 7	0	28	1.18	458	4	129	69	0.3	0.7	0 7	729
1094	1000	0 8	0	30	1.32	510	5	127	69	0.3	0.7	0 8	720
1203	1100	0 9	0	31	1.46	562	5	125	69	0.4	0.8	0 9	710
1312	1200	0 9	0	33	1.60	614	6	122	69	0.4	0.9	0 10	700
1422	1300	0 10	0	35	1.74	667	7	120	69	0.5	1.0	0 11	691
1531	1400	0 11	0	36	1.89	720	8	118	69	0.5	1.0	0 12	681
1640	1500	0 12	0	38	2.04	773	9	116	69	0.5	1.1	0 13	672

*Numbers in column 2 are in degrees, and in column 5 in 1/16 degrees.

GERMAN 88-MM ANTIAIRCRAFT GUN MATERIEL

Range		Quadrant Elevation	Drift	For S/30 Fuze Only Fuze Setting at 0 Meters Remaining Flight	Time of Flight	Maximum Ordinate		1/16 deg Changes the Point of Impact by	Means (50 percent)			Angle of Impact	Velocity
						Distance	Height		Range	Dispersion Width	Height		
yd	m	deg	mils	deg from cross	sec	m	m	m	m	m	m	deg	m/s
	1	2*	3	4	5	6	7	8	9	10	11	12*	13
1750	1600	0 13	0	40	2.19	826	10	114	69	0.6	1.2	0 14	663
1859	1700	0 14	1	41	2.34	879	11	111	69	0.6	1.2	0 15	653
1968	1800	0 14	1	43	2.49	933	12	109	69	0.7	1.3	1 1	644
2078	1900	0 15	1	45	2.65	987	13	107	69	0.7	1.4	1 2	635
2187	2000	1 0	1	47	2.81	1041	14	105	70	0.8	1.5	1 3	626
2297	2100	1 1	1	48	2.97	1095	15	103	70	0.8	1.6	1 5	618
2406	2200	1 2	1	50	3.13	1150	16	101	70	0.9	1.7	1 6	609
2515	2300	1 3	1	52	3.29	1205	17	99	70	0.9	1.9	1 7	600
2625	2400	1 4	1	54	3.46	1260	19	97	70	0.9	2.0	1 9	592
2734	2500	1 5	1	56	3.63	1316	20	95	70	1.0	2.1	1 10	583
2843	2600	1 6*	1	58	3.80	1372	21	93	70	1.0	2.2	1 12*	575
2953	2700	1 7	1	60	3.97	1429	23	91	70	1.1	2.3	1 14	567
3062	2800	1 8	1	62	4.15	1486	24	89	70	1.1	2.4	1 15	558
3172	2900	1 10	1	64	4.33	1543	26	87	70	1.1	2.6	2 1	550
3281	3000	1 11	1	66	4.51	1601	28	85	70	1.2	2.7	2 3	542
3390	3100	1 12	1	68	4.70	1659	30	83	70	1.2	2.8	2 5	534
3500	3200	1 13	1	70	4.89	1717	32	81	70	1.3	3.0	2 7	526
3609	3300	1 14	1	72	5.08	1775	35	79	70	1.3	3.1	2 9	518
3718	3400	2 0	1	75	5.28	1834	37	78	70	1.4	3.3	2 11	510
3828	3500	2 1	1	77	5.48	1893	40	76	70	1.4	3.5	2 13	503

FIRING TABLES

3937	3600	2 2	1	79	5.68	1952	42	74	70	1.5	3.7	2 15	495
4046	3700	2 4	2	82	5.89	2011	45	73	71	1.5	3.9	3 2	487
4156	3800	2 5	2	84	6.10	2071	48	71	71	1.6	4.1	3 4	479
4265	3900	2 6	2	86	6.31	2131	51	70	71	1.6	4.3	3 7	472
4374	4000	2 8	2	89	6.52	2191	54	68	71	1.7	4.6	3 10	464
4484	4100	2 9	2	91	6.74	2252	58	67	71	1.7	4.8	3 13	457
4593	4200	2 11	2	94	6.96	2313	61	65	71	1.8	5.1	4 0	450
4703	4300	2 12	2	96	7.18	2374	65	64	71	1.8	5.3	4 3	442
4812	4400	2 14	2	99	7.41	2436	69	62	71	1.9	5.6	4 6	435
4921	4500	3 0	2	101	7.64	2498	73	61	71	1.9	5.9	4 10	428
5031	4600	3 1	2	104	7.87	2560	77	60	71	2.0	6.1	4 13	421
5140	4700	3 3	2	107	8.11	2622	81	58	71	2.0	6.4	5 1	414
5249	4800	3 5	2	109	8.35	2685	86	57	71	2.1	6.7	5 4	407
5359	4900	3 6	2	112	8.60	2748	91	56	71	2.1	7.0	5 8	400
5468	5000	3 8	2	115	8.85	2811	96	55	71	2.2	7.3	5 12	393
5577	5100	3 10*	3	118	9.11	2874	102	53	71	2.3	7.6	6 1*	387
5687	5200	3 12	3	121	9.37	2937	108	52	71	2.3	8.0	6 5	381
5796	5300	3 14	3	124	9.64	3001	114	51	72	2.4	8.4	6 10	375
5905	5400	4 0	3	127	9.91	3065	121	50	72	2.5	8.8	6 14	369
6015	5500	4 2	3	130	10.18	3129	128	49	72	2.5	9.2	7 3	363
6124	5600	4 4	3	133	10.46	3193	135	48	72	2.6	9.7	7 8	358
6234	5700	4 6	3	136	10.74	3257	143	47	72	2.7	10	7 13	353
6343	5800	4 8	3	140	11.03	3321	151	46	72	2.7	10	8 2	348
6452	5900	4 11	3	143	11.32	3385	159	45	72	2.8	11	8 7	344
6562	6000	4 13	3	146	11.62	3450	168	44	72	2.9	11	8 12	340

*Numbers in column 2 are in degrees, and in column 5 in 1/16 degrees.

GERMAN 88-MM ANTIAIRCRAFT GUN MATERIEL

Range		Quadrant Elevation	Drift	For S/30 Fuze Only Fuze Setting at 0 Meters Remaining Flight	Time of Flight	Maximum Ordinate		1/16 deg Changes the Point of Impact by	Means (50 percent) Dispersion			Angle of Impact	Velocity
						Distance	Height		Range	Width	Height		
yd	m	deg	mils	deg from cross	sec	m	m	m	m	m	m	deg	m/s
1		2*	3	4	5	6	7	8	9	10	11	12*	13
6671	6100	4 15	3	150	11.92	3515	177	43	72	3.0	12	9 1	336
6780	6200	5 2	3	153	12.22	3580	187	42	72	3.0	12	9 6	332
6890	6300	5 4	4	157	12.53	3645	197	41	72	3.1	13	9 12	329
6999	6400	5 7	4	160	12.84	3710	207	40	72	3.2	13	10 1	326
7108	6500	5 9	4	164	13.15	3775	218	39	72	3.3	14	10 7	323
7218	6600	5 12	4	167	13.47	3840	229	39	73	3.4	14	10 13	320
7327	6700	5 15	4	171	13.79	3905	240	38	73	3.4	15	11 3	317
7437	6800	6 1	4	175	14.11	3970	252	37	73	3.5	15	11 9	314
7546	6900	6 4	4	178	14.44	4035	264	36	73	3.6	16	11 15	312
7655	7000	6 7	4	182	14.77	4100	276	35	73	3.7	16	12 5	309
7765	7100	6 10	4	186	15.10	4165	289	35	73	3.8	17	12 12	307
7874	7200	6 13	4	190	15.43	4230	302	34	73	3.9	17	13 2	305
7983	7300	7 0	5	193	15.77	4295	316	33	73	4.0	18	13 9	303
8093	7400	7 3	5	197	16.11	4360	330	33	73	4.1	18	14	301
8202	7500	7 6	5	201	16.45	4424	345	32	73	4.2	19	14 7	299
8311	7600	7 9*	5	205	16.80	4488	360	32	73	4.3	19	14 14*	297
8421	7700	7 12	5	209	17.15	4552	376	31	73	4.4	20	15 5	295
8530	7800	7 15	5	213	17.50	4616	393	30	73	4.5	21	15 13	293
8640	7900	8 3	5	217	17.86	4680	410	30	74	4.6	21	16 4	291
8749	8000	8 6	5	221	18.22	4744	428	29	74	4.7	22	16 12	289

FIRING TABLES

8858	8100	8 10	6	225	18.59	4808	446	29	74	4.8	23	17 3	287
8968	8200	8 13	6	230	18.96	4871	465	28	74	4.9	24	17 11	286
9077	8300	9 1	6	234	19.33	4934	485	28	74	5.0	24	18 3	284
9186	8400	9 5	6	238	19.71	4997	505	27	74	5.1	25	18 11	282
9296	8500	9 8	6	242	20.09	5060	526	27	74	5.2	26	19 3	281
9405	8600	9 12	6	247	20.47	5123	548	26	74	5.3	27	19 11	279
9514	8700	10 0	6	251	20.86	5185	570	26	74	5.4	27	20 4	278
9624	8800	10 4	6	255	21.25	5247	593	25	74	5.6	28	20 12	276
9733	8900	10 8	7	260	21.64	5309	616	25	74	5.7	29	21 5	275
9842	9000	10 12	7	264	22.04	5371	640	25	75	5.8	30	21 13	273
9952	9100	11 0	7	269	22.44	5432	665	24	75	5.9	31	22 6	272
10061	9200	11 4	7	274	22.85	5493	690	24	75	6.1	31	22 14	271
10171	9300	11 9	7	278	23.26	5554	716	23	75	6.2	32	23 7	269
10280	9400	11 13	7	283	23.67	5615	743	23	75	6.3	33	24 0	268
10389	9500	12 2	8	288	24.09	5676	770	23	75	6.5	34	24 9	267
10499	9600	12 6	8	292	24.51	5737	798	22	76	6.6	35	25 2	266
10608	9700	12 11	8	297	24.93	5797	827	22	76	6.7	36	25 11	265
10717	9800	12 15	8	302	25.36	5857	857	22	76	6.9	37	26 4	264
10827	9900	13 4	8	307	25.79	5917	887	21	76	7.0	38	26 14	263
10936	10000	13 9	8	312	26.22	5977	918	21	76	7.2	39	27 7	262
11045	10100	13 13	9	317	26.66	6037	950	21	77	7.3	41	28 0	261
11155	10200	14 2	9	322	27.10	6097	983	20	77	7.5	42	28 10	260
11264	10300	14 7	9	327	27.55	6156	1016	20	77	7.6	43	29 3	259
11374	10400	14 12	9	332	28.01	6215	1050	20	77	7.8	44	29 13	259
11483	10500	15 1	9	337	28.46	6274	1085	19	78	7.9	45	30 6	258

*Numbers in column 2 are in degrees, and in column 5 in 1/16 degrees.

GERMAN 88-MM ANTIAIRCRAFT GUN MATERIEL

Range		Quadrant Elevation	Drift	For S/30 Fuze Only Fuze Setting at 0 Meters Remaining Flight	Time of Flight	Maximum Ordinate		1/16 deg Changes the Point of Impact by	Means (50 percent)			Angle of Impact	Velocity
						Distance	Height		Range	Dispersion Width	Height		
yd	m	deg	mils	deg from cross	sec	m	m	m	m	m	m	deg	m/s
1	1	2*	3	4	5	6	7	8	9	10	11	12*	13
11592	10600	15 6	10	—	28.92	6333	1121	19	78	8.1	47	31 0	257
11702	10700	15 12	10	—	29.39	6392	1158	19	78	8.3	48	31 10	257
11811	10800	16 1	10	—	29.87	6450	1196	19	79	8.4	49	32 3	256
11920	10900	16 7	10	—	30.35	6508	1235	18	79	8.6	51	32 13	256
12030	11000	16 12	10	—	30.84	6566	1275	18	79	8.8	52	33 7	255
12139	11100	17 2	11	—	31.34	6624	1316	18	80	9.0	54	34 1	255
12248	11200	17 7	11	—	31.84	6681	1358	18	80	9.1	55	34 11	255
12358	11300	17 13	11	—	32.35	6738	1401	17	81	9.3	57	35 5	254
12467	11400	18 3	11	—	32.87	6795	1445	17	81	9.5	59	35 15	254
12577	11500	18 9	12	—	33.39	6852	1490	17	82	9.7	61	36 9	254
12686	11600	18 15	12	—	33.92	6908	1537	16	82	9.9	62	37 3	254
12795	11700	19 5	12	—	34.46	6964	1585	16	83	10	64	37 13	254
12905	11800	19 12	13	—	35.00	7020	1635	16	84	10	66	38 7	254
13014	11900	20 2	13	—	35.55	7076	1686	16	84	11	68	39 1	253
13123	12000	20 9	13	—	36.11	7131	1739	15	85	11	70	39 11	253
13233	12100	20 15	13	—	36.67	7186	1794	15	86	11	73	40 5	254
13342	12200	21 6	14	—	37.24	7241	1850	15	86	11	75	41 0	254
13451	12300	21 13	14	—	37.82	7296	1908	14	87	11	77	41 10	254
13561	12400	22 4	14	—	38.40	7351	1968	14	88	12	80	42 4	254
13670	12500	22 11	15	—	38.99	7406	2029	14	89	12	83	42 15	254

FIRING TABLES

13779	12600	23 3	15	—	39.59	7461	2092	14	90	12	83	43 9	254
13889	12700	23 10	15	—	40.20	7516	2157	13	91	12	85	44 4	255
13998	12800	24 2	16	—	40.82	7571	2224	13	92	12	91	44 14	255
14108	12900	24 10	16	—	41.45	7625	2293	13	93	13	94	45 9	255
14217	13000	25 2	16	—	42.09	7679	2364	12	94	13	98	46 3	256
14326	13100	25 10	17	—	42.74	7733	2437	12	95	13	101	46 14	256
14436	13200	26 3	17	—	43.41	7787	2513	12	96	13	105	47 9	257
14545	13300	26 12	18	—	44.10	7841	2592	11	97	14	109	48 4	257
14654	13400	27 5	18	—	44.81	7895	2674	11	99	14	114	48 15	258
14764	13500	27 14	19	—	45.54	7949	2760	10	100	14	118	49 10	259
14873	13600	28 8	19	—	46.29	8003	2850	10	102	14	123	50 6	260
14982	13700	29 2	20	—	47.06	8056	2945	10	103	15	128	51 2	261
15092	13800	29 13	20	—	47.86	8109	3045	9	105	15	134	51 14	262
15201	13900	30 8	21	—	48.69	8162	3151	9	107	15	140	52 10	263
15311	14000	31 4	21	—	49.56	8215	3264	8	109	16	147	53 7	264
15420	14100	32 0	22	—	50.47	8268	3385	8	111	16	154	54 4	265
15529	14200	32 14	22	—	51.43	8321	3515	7	113	16	162	55 1	267
15639	14300	33 12	23	—	52.45	8374	3655	7	115	17	171	55 15	268
15748	14400	34 11	23	—	53.54	8427	3806	6	118	17	181	56 14	270
15857	14500	35 12	24	—	54.76	8480	3970	6	121	18	193	57 13	272
15967	14600	36 15	25	—	56.13	8531	4158	5	125	18	206	58 14	274
16076	14700	38 6	26	—	57.76	8583	4387	4	129	19	222	60 2	276
16185	14800	40 4	27	—	59.87	8613	4692	2	135	20	245	61 11	280

*Numbers in column 2 are in degrees, and in column 5 in 1/16 degrees.

GERMAN 88-MM ANTIAIRCRAFT GUN MATERIEL

89. FIRING TABLE FOR THE 8.8 cm FLAK 18 AND 36 WITH 8.8 cm A.P. SHELL WITH BASE FUZE (TABLE II).

Table II

Muzzle velocity = 810 m/s or 2657 ft/sec   Weight of shell 9.65 kg or 20.75 lb
Weight of air at ground level = 1.22 kg/cu m or 533 grains/cu ft

Range	Quadrant Elevation	Drift	Time of Flight	Angle of Impact	Terminal Velocity	Means (50 percent) Dispersion Height	Width
m	deg	mils	sec	deg	m/s	m	m
1	2*	3	4	5	6	7	8
—	—	—	—	—	810	—	—
100	0 1	0	0.12	0 1*	800	—	0.1
200	0 1	0	0.25	0 2	790	0.2	0.1
300	0 2	0	0.37	0 2	780	0.2	0.2
400	0 3	0	0.49	0 3	770	0.3	0.2
500	0 3	0	0.62	0 3	761	0.3	0.2
600	0 4	0	0.74	0 4	752	0.4	0.2
700	0 5	0	0.86	0 5	742	0.5	0.3
800	0 6	0	1.00	0 6	733	0.5	0.3
900	0 6	0	1.12	0 7	725	0.6	0.3
1000	0 7	0	1.25	0 8	716	0.7	0.4
1100	0 8	0	1.39	0 9	708	0.7	0.4
1200	0 9	0	1.52	0 10	700	0.8	0.5
1300	0 10	0	1.65	0 11	691	0.9	0.5
1400	0 11	0	1.80	0 12	683	1.0	0.5
1500	0 12	0	1.94	0 13	676	1.1	0.6

FIRING TABLES

Range								
1600	0	12	0	2.08	0	14	668	1.2 0.6
1700	0	13	0	2.23	0	15	660	1.3 0.7
1800	0	14	1	2.37	1	0	652	1.4 0.7
1900	0	15	1	2.52	1	1	645	1.5 0.7
2000	1	0	1	2.68	1	2	637	1.6 0.8
2100	1	1	1	2.83	1	3	629	1.6 0.8
2200	1	2	1	2.98	1	5	623	1.8 0.9
2300	1	3	1	3.13	1	6	615	1.9 0.9
2400	1	4	1	3.30	1	7	608	2.0 0.9
2500	1	5	1	3.46	1	8	602	2.1 1.0
2600	1	6	1	3.62	1	9	595	2.2 1.0
2700	1	7	1	3.79	1	10	588	2.3 1.1
2800	1	8	1	3.96	1	11	581	2.4 1.1
2900	1	9	1	4.13	1	13	575	2.6 1.2
3000	1	10	1	4.31	1	15	569	2.7 1.2
3100	1	11	1	4.49	2	1	562	2.9 1.2
3200	1	12	1	4.67	2	3	557	3.1 1.3
3300	1	13	1	4.86	2	5	551	3.3 1.3
3400	1	15	1	5.05	2	7	544	3.5 1.4
3500	2	0	1	5.24	2	9	539	3.7 1.4
3600	2	1	1	5.44	2	11	533	3.9 1.5
3700	2	2	1	5.64	2	13	527	4.2 1.5
3800	2	3	1	5.84	2	15	521	4.4 1.6
3900	2	5	2	6.05	3	1	516	4.7 1.6
4000	2	6	2	6.25	3	3	510	5.0 1.7

*Numbers in column 2 are in degrees, and in column 5 in 1/16 degrees.

GERMAN 88-MM ANTIAIRCRAFT GUN MATERIEL

90. CONVERSION TABLES.

Yards to Meters
1 yard = 0.91440183 meters

YARDS	0	10	20	30	40	50	60	70	80	90
0	.00	9.14	18.29	27.43	36.58	45.72	54.86	64.01	73.15	82.30
100	91.44	100.58	109.73	118.87	128.02	137.16	146.30	155.45	164.59	173.74
200	182.88	192.02	201.17	210.31	219.46	228.60	237.74	246.89	256.03	265.18
300	274.32	283.46	292.61	301.75	310.90	320.04	329.18	338.33	347.47	356.62
400	365.76	374.90	384.05	393.19	402.34	411.48	420.62	429.77	438.91	448.06
500	457.20	466.34	475.49	484.63	493.78	502.92	512.07	521.21	530.35	539.50
600	548.64	557.79	566.93	576.07	585.22	594.36	603.51	612.65	621.79	630.94
700	640.06	649.23	658.37	667.51	676.66	685.80	694.95	704.09	713.23	722.38
800	731.52	740.67	749.81	758.95	768.10	777.24	786.39	795.53	804.67	813.82
900	822.96	832.11	841.25	850.39	859.54	868.68	877.83	886.97	896.11	905.26
1000	914.40	923.55	932.69	941.83	950.98	960.12	969.27	978.41	987.55	996.70

Meters to Yards
1 meter = 1.0936111 yards

METERS	0	10	20	30	40	50	60	70	80	90
0	.00	10.94	21.87	32.81	43.74	54.68	65.62	76.55	87.49	98.42
100	109.36	120.30	131.23	142.17	153.11	164.04	174.98	185.91	196.85	207.79
200	218.72	229.66	240.59	251.53	262.47	273.40	284.34	295.27	306.21	317.15
300	328.08	339.02	349.96	360.89	371.83	382.76	393.70	404.64	415.57	426.51
400	437.44	448.38	459.32	470.25	481.19	492.12	503.06	514.00	524.93	535.87
500	546.81	557.74	568.68	579.61	590.55	601.49	612.42	623.36	634.29	645.23
600	656.18	667.10	678.04	688.97	699.91	710.85	721.78	732.72	743.66	754.59
700	765.53	776.46	787.04	798.34	809.27	820.21	831.14	842.08	853.02	863.95
800	874.89	885.82	896.76	907.70	918.63	929.57	940.51	951.44	962.38	973.31
900	984.25	995.19	1006.12	1017.06	1027.99	1038.93	1049.87	1060.80	1071.74	1082.67
1000	1093.61	1104.55	1115.48	1126.42	1137.36	1148.29	1159.23	1170.16	1181.10	1192.04

FIRING TABLES

Angular Conversion Table — Degrees to Mils

Degrees	0	1	2	3	4	5	6	7	8	9
00	0	18	36	53	71	89	107	124	142	160
10	178	196	213	231	249	267	284	302	320	338
20	356	373	391	409	427	444	462	480	498	516
30	533	551	569	587	604	622	640	658	676	693
40	711	729	747	764	782	800	818	836	853	871
50	889	907	924	942	960	978	996	1013	1031	1049
60	1067	1084	1102	1120	1138	1156	1173	1191	1209	1227
70	1244	1262	1280	1298	1316	1333	1351	1369	1387	1404
80	1422	1440	1458	1476	1493	1511	1529	1547	1564	1582
90	1600									

(Conversion Factor, 1 deg = 17.77778 mils)

GERMAN 88-MM ANTIAIRCRAFT GUN MATERIEL

CHAPTER 6

REFERENCES

91. STANDARD NOMENCLATURE LISTS.

   a. Cleaning, preserving and lubricating materials; recoil fluids, special oils, and miscellaneous related items .... SNL K-1

92. EXPLANATORY PUBLICATIONS.

   a. Ammunition, general .... TM 9-1900

   b. Chemical decontamination materials and equipment .... TM 3-220

   c. Cleaning, preserving, lubricating, and welding materials and similar items issued by the Ordnance Department .... TM 9-850

   d. Defense against chemical attack.......................... FM 21-40

   e. Product guide .... OFSB 6-2

INDEX

INDEX

INDEX

$\left[\begin{array}{l}\text{A.G. 300.7 (6 Jul. 1943)}\\\text{O.O. 461/ Raritan Arsenal (3 Jul. 1943) (R)}\end{array}\right]$

BY ORDER OF THE SECRETARY OF WAR:

G. C. MARSHALL,
Chief of Staff.

OFFICIAL:
J. A. ULIO,
Major General,
The Adjutant General.

DISTRIBUTION: **X**

(For explanation of symbols, see FM 21-6)

Printed in Great Britain
by Amazon